Take a Chance

Britannica

ENCYCLOPÆDIA BRITANNICA EDUCATIONAL CORPORATION

Mathematics in Context is a comprehensive middle grades curriculum. It was developed in collaboration with the Wisconsin Center for Education Research, School of Education, University of Wisconsin–Madison and the Freudenthal Institute at the University of Utrecht, The Netherlands, with the support of National Science Foundation Grant No. 9054928.

National Science Foundation

Opinions expressed are those of the authors
and not necessarily those of the Foundation

ISBN 0-7826-1497-3
1 2 3 4 5 6 7 8 9 10   99 98 97

The *Mathematics in Context* Development Team

Mathematics in Context is a comprehensive middle grades curriculum. The National Science Foundation funded the National Center for Research in Mathematical Sciences Education at the University of Wisconsin–Madison to develop and field-test the materials from 1991 through 1996. The Freudenthal Institute at the University of Utrecht in The Netherlands is the main subcontractor responsible for the development of the student and assessment materials.

The initial version of *Take a Chance* was developed by Vincent Jonker, Frans van Galen, Nina Boswinkel, and Monica Wijers. It was adapted for use in American schools by Aaron N. Simon, Gail Burrill, and James A. Middleton.

National Center for Research in Mathematical Sciences Education Staff

Thomas A. Romberg
Director

Joan Daniels Pedro
Assistant to the Director

Gail Burrill
Coordinator
Field Test Materials

Margaret R. Meyer
Coordinator
Pilot Test Materials

Mary Ann Fix
Editorial Coordinator

Sherian Foster
Editorial Coordinator

James A. Middleton
Pilot Test Coordinator

Project Staff

Jonathan Brendefur
Laura J. Brinker
James Browne
Jack Burrill
Rose Byrd
Peter Christiansen
Barbara Clarke
Doug Clarke
Beth R. Cole

Fae Dremock
Jasmina Milinkovic
Margaret A. Pligge
Mary C. Shafer
Julia A. Shew
Aaron N. Simon
Marvin Smith
Stephanie Z. Smith
Mary S. Spence

Freudenthal Institute Staff

Jan de Lange
Director

Els Feijs
Coordinator

Martin van Reeuwijk
Coordinator

Project Staff

Mieke Abels
Nina Boswinkel
Frans van Galen
Koeno Gravemeijer
Marja van den Heuvel-Panhuizen
Jan Auke de Jong
Vincent Jonker
Ronald Keijzer

Martin Kindt
Jansie Niehaus
Nanda Querelle
Anton Roodhardt
Leen Streefland
Adri Treffers
Monica Wijers
Astrid de Wild

Acknowledgments

Several school districts used and evaluated one or more versions of the materials: Ames Community School District, Ames, Iowa; Parkway School District, Chesterfield, Missouri; Stoughton Area School District, Stoughton, Wisconsin; Madison Metropolitan School District, Madison, Wisconsin; Milwaukee Public Schools, Milwaukee, Wisconsin; and Dodgeville School District, Dodgeville, Wisconsin. Two sites were involved in staff development as well as formative evaluation of materials: Culver City, California, and Memphis, Tennessee. Two sites were developed through partnership with Encyclopædia Britannica Educational Corporation: Miami, Florida, and Puerto Rico. University Partnerships were developed with mathematics educators who worked with preservice teachers to familiarize them with the curriculum and to obtain their advice on the curriculum materials. The materials were also used at several other schools throughout the United States.

We at Encyclopædia Britannica Educational Corporation extend our thanks to all who had a part in making this program a success. Some of the participants instrumental in the program's development are as follows:

Allapattah Middle School
Miami, Florida
Nemtalla (Nikolai) Barakat

Ames Middle School
Ames, Iowa
Kathleen Coe
Judd Freeman
Gary W. Schnieder
Ronald H. Stromen
Lyn Terrill

Bellerive Elementary
Creve Coeur, Missouri
Judy Hetterscheidt
Donna Lohman
Gary Alan Nunn
Jakke Tchang

Brookline Public Schools
Brookline, Massachusetts
Rhonda K. Weinstein
Deborah Winkler

Cass Middle School
Milwaukee, Wisconsin
Tami Molenda
Kyle F. Witty

Central Middle School
Waukesha, Wisconsin
Nancy Reese

Craigmont Middle School
Memphis, Tennessee
Sharon G. Ritz
Mardest K. VanHooks

Crestwood Elementary
Madison, Wisconsin
Diane Hein
John Kalson

Culver City Middle School
Culver City, California
Marilyn Culbertson
Joel Evans
Joy Ellen Kitzmiller
Patricia R. O'Connor
Myrna Ann Perks, Ph.D.
David H. Sanchez
John Tobias
Kelley Wilcox

Cutler Ridge Middle School
Miami, Florida
Lorraine A. Valladares

Dodgeville Middle School
Dodgeville, Wisconsin
Jacqueline A. Kamps
Carol Wolf

Edwards Elementary
Ames, Iowa
Diana Schmidt

Fox Prairie Elementary
Stoughton, Wisconsin
Tony Hjelle

Grahamwood Elementary
Memphis, Tennessee
M. Lynn McGoff
Alberta Sullivan

Henry M. Flagler Elementary
Miami, Florida
Frances R. Harmon

Horning Middle School
Waukesha, Wisconsin
Connie J. Marose
Thomas F. Clark

Huegel Elementary
Madison, Wisconsin
Nancy Brill
Teri Hedges
Carol Murphy

Hutchison Middle School
Memphis, Tennessee
Maria M. Burke
Vicki Fisher
Nancy D. Robinson

Idlewild Elementary
Memphis, Tennessee
Linda Eller

Jefferson Elementary
Santa Ana, California
Lydia Romero-Cruz

Jefferson Middle School
Madison, Wisconsin
Jane A. Beebe
Catherine Buege
Linda Grimmer
John Grueneberg
Nancy Howard
Annette Porter
Stephen H. Sprague
Dan Takkunen
Michael J. Vena

Jesus Sanabria Cruz School
Yabucoa, Puerto Rico
Andreíta Santiago Serrano

John Muir Elementary School
Madison, Wisconsin
Julie D'Onofrio
Jane M. Allen-Jauch
Kent Wells

Kegonsa Elementary
Stoughton, Wisconsin
Mary Buchholz
Louisa Havlik
Joan Olsen
Dominic Weisse

Linwood Howe Elementary
Culver City, California
Sandra Checel
Ellen Thireos

Mitchell Elementary
Ames, Iowa
Henry Gray
Matt Ludwig

New School of Northern Virginia
Fairfax, Virginia
Denise Jones

Northwood Elementary
Ames, Iowa
Eleanor M. Thomas

Orchard Ridge Elementary
Madison, Wisconsin
Mary Paquette
Carrie Valentine

Parkway West Middle School
Chesterfield, Missouri
Elissa Aiken
Ann Brenner
Gail R. Smith

Ridgeway Elementary
Ridgeway, Wisconsin
Lois Powell
Florence M. Wasley

Roosevelt Elementary
Ames, Iowa
Linda A. Carver

Roosevelt Middle
Milwaukee, Wisconsin
Sandra Simmons

Ross Elementary
Creve Coeur, Missouri
Annette Isselhard
Sheldon B. Korklan
Victoria Linn
Kathy Stamer

St. Joseph's School
Dodgeville, Wisconsin
Rita Van Dyck
Sharon Wimer

St. Maarten Academy
St. Peters, St. Maarten, NA
Shareed Hussain

Sarah Scott Middle School
Milwaukee, Wisconsin
Kevin Haddon

Sawyer Elementary
Ames, Iowa
Karen Bush Hoiberg

Sennett Middle School
Madison, Wisconsin
Brenda Abitz
Lois Bell
Shawn M. Jacobs

Sholes Middle School
Milwaukee, Wisconsin
Chris Gardner
Ken Haddon

Stephens Elementary
Madison, Wisconsin
Katherine Hogan
Shirley M. Steinbach
Kathleen H. Vegter

Stoughton Middle School
Stoughton, Wisconsin
Sally Bertelson
Polly Goepfert
Jacqueline M. Harris
Penny Vodak

Toki Middle School
Madison, Wisconsin
Gail J. Anderson
Vicky Grice
Mary M. Ihlenfeldt
Steve Jernegan
Jim Leidel
Theresa Loehr
Maryann Stephenson
Barbara Takkunen
Carol Welsch

Trowbridge Elementary
Milwaukee, Wisconsin
Jacqueline A. Nowak

W. R. Thomas Middle School
Miami, Florida
Michael Paloger

Wooddale Elementary Middle School
Memphis, Tennessee
Velma Quinn Hodges
Jacqueline Marie Hunt

Yahara Elementary
Stoughton, Wisconsin
Mary Bennett
Kevin Wright

Site Coordinators

Mary L. Delagardelle—Ames Community Schools, Ames, Iowa

Dr. Hector Hirigoyen—Miami, Florida

Audrey Jackson—Parkway School District, Chesterfield, Missouri

Jorge M. López—Puerto Rico

Susan Militello—Memphis, Tennessee

Carol Pudlin—Culver City, California

Reviewers and Consultants

Michael N. Bleicher
Professor of Mathematics
University of Wisconsin–Madison
Madison, WI

Diane J. Briars
Mathematics Specialist
Pittsburgh Public Schools
Pittsburgh, PA

Donald Chambers
Director of Dissemination
University of Wisconsin–Madison
Madison, WI

Don W. Collins
Assistant Professor of Mathematics Education
Western Kentucky University
Bowling Green, KY

Joan Elder
Mathematics Consultant
Los Angeles Unified School District
Los Angeles, CA

Elizabeth Fennema
Professor of Curriculum and Instruction
University of Wisconsin-Madison
Madison, WI

Nancy N. Gates
University of Memphis
Memphis, TN

Jane Donnelly Gawronski
Superintendent
Escondido Union High School
Escondido, CA

M. Elizabeth Graue
Assistant Professor of Curriculum and Instruction
University of Wisconsin–Madison
Madison, WI

Jodean E. Grunow
Consultant
Wisconsin Department of Public Instruction
Madison, WI

John G. Harvey
Professor of Mathematics and Curriculum & Instruction
University of Wisconsin–Madison
Madison, WI

Simon Hellerstein
Professor of Mathematics
University of Wisconsin–Madison
Madison, WI

Elaine J. Hutchinson
Senior Lecturer
University of Wisconsin–Stevens Point
Stevens Point, WI

Richard A. Johnson
Professor of Statistics
University of Wisconsin–Madison
Madison, WI

James J. Kaput
Professor of Mathematics
University of Massachusetts–Dartmouth
Dartmouth, MA

Richard Lehrer
Professor of Educational Psychology
University of Wisconsin–Madison
Madison, WI

Richard Lesh
Professor of Mathematics
University of Massachusetts–Dartmouth
Dartmouth, MA

Mary M. Lindquist
Callaway Professor of Mathematics Education
Columbus College
Columbus, GA

Baudilio (Bob) Mora
Coordinator of Mathematics & Instructional Technology
Carrollton-Farmers Branch Independent School District
Carrollton, TX

Paul Trafton
Professor of Mathematics
University of Northern Iowa
Cedar Falls, IA

Norman L. Webb
Research Scientist
University of Wisconsin–Madison
Madison, WI

Paul H. Williams
Professor of Plant Pathology
University of Wisconsin–Madison
Madison, WI

Linda Dager Wilson
Assistant Professor
University of Delaware
Newark, DE

Robert L. Wilson
Professor of Mathematics
University of Wisconsin–Madison
Madison, WI

TABLE OF CONTENTS

BRITANNICA
Mathematics in Context

Dear Teacher,

Welcome! *Mathematics in Context* is designed to reflect the National Council of Teachers of Mathematics Standards for School Mathematics and to ground mathematical content in a variety of real-world contexts. Rather than relying on you to explain and demonstrate generalized definitions, rules, or algorithms, students investigate questions directly related to a particular context and construct mathematical understanding and meaning from that context.

The curriculum encompasses 10 units per grade level. This unit is designed to be the second in the statistics strand for grade 5/6, but it also lends itself to independent use—to introduce students to elementary notions of probability using spinners, number cubes, coins, and other tools that can represent chance situations.

In addition to the Teacher Guide and Student Books, *Mathematics in Context* offers the following components that will inform and support your teaching:

- *Teacher Resource and Implementation Guide,* which provides an overview of the complete system, including program implementation, philosophy, and rationale

- *Number Tools,* which is a series of blackline masters that serve as review sheets or practice pages involving number issues and basic skills

- *News in Numbers,* which is a set of additional activities that can be inserted between or within other units; it includes a number of measurement problems that require estimation.

- *Teacher Preparation Videos,* which present comprehensive overviews of the units to help with lesson preparation

Thank you for choosing *Mathematics in Context.* We wish you success and inspiration!

Sincerely,

The Mathematics in Context Development Team

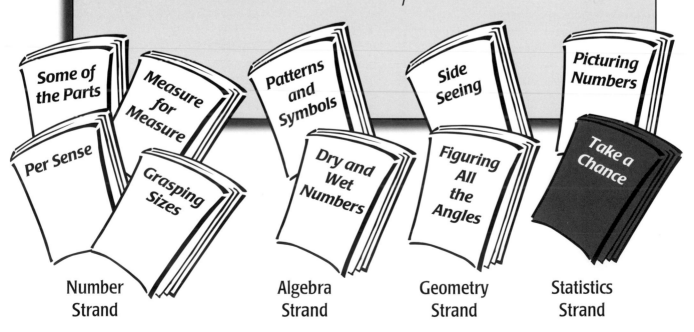

Some of the Parts

Per Sense

Measure for Measure

Grasping Sizes

Patterns and Symbols

Dry and Wet Numbers

Side Seeing

Figuring All the Angles

Picturing Numbers

Take a Chance

Number Strand

Algebra Strand

Geometry Strand

Statistics Strand

Overview

B R I T A N N I C A

Mathematics in Context

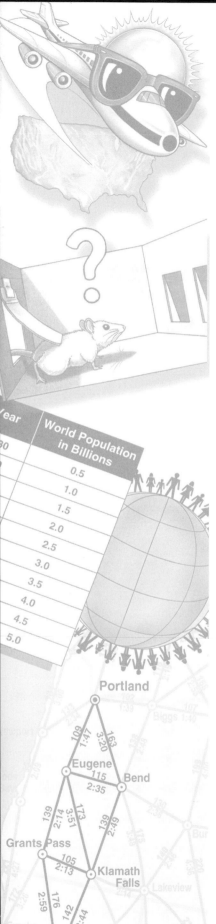

How to Use This Book

This unit is one of 40 for the middle grades. Each unit can be used independently; however, the 40 units are designed to make up a complete, connected curriculum (10 units per grade level). There is a Student Book and a Teacher Guide for each unit.

Each Teacher Guide comprises elements that assist the teacher in the presentation of concepts and in understanding the general direction of the unit and the program as a whole. Becoming familiar with this structure will make using the units easier.

Each Teacher Guide consists of six basic parts:

- Overview
- Student Material and Teaching Notes
- Assessment Activities and Solutions
- Glossary
- Blackline Masters
- Try This! Solutions

Overview

Before beginning this unit, read the Overview in order to understand the purpose of the unit and to develop strategies for facilitating instruction. The Overview provides helpful information about the unit's focus, pacing, goals, and assessment, as well as explanations about how the unit fits with the rest of the *Mathematics in Context* curriculum.

Note: After reading the Overview, view the Teacher Preparation Videotape that corresponds with the strand. The video models several activities from the strand.

Student Materials and Teaching Notes

This Teacher Guide contains all of the student pages (except the Try This! activities), each of which faces a page of solutions, samples of students' work, and hints and comments about how to facilitate instruction. Note: Solutions for the Try This! activities can be found at the back of this Teacher Guide.

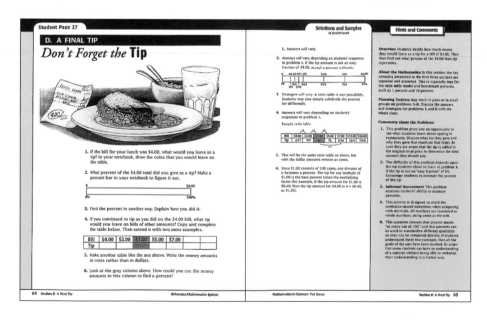

Each section within the unit begins with a two-page spread that describes the work students do, the goals of the section, new vocabulary, and materials needed, as well as providing information about the mathematics in the section and ideas for pacing, planning instruction, homework, and assessment.

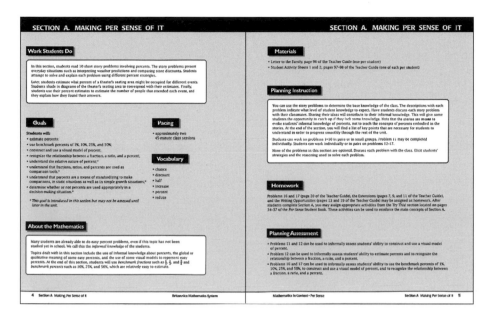

Assessment Activities and Solutions

Information about assessment can be found in several places in this Teacher Guide. General information about assessment is given in the Overview; informal assessment opportunities are identified on the teacher pages that face each student page; and the Assessment Activities section of this guide provides formal assessment opportunities.

Glossary

The Glossary defines all vocabulary words listed on the Section Opener pages. It includes mathematical terms that may be new to students, as well as words associated with the contexts introduced in the unit. (Note: The Student Book does not have a glossary. This allows students to construct their own definitions, based on their personal experiences with the unit activities.)

Blackline Masters

At the back of this Teacher Guide are blackline masters for photocopying. The blackline masters include a letter to families (to be sent home with students before beginning the unit), several student activity sheets, and assessment masters.

Try This! Solutions

Also included in the back of this Teacher Guide are the solutions to several Try This! activities—one related to each section of the unit—that can be used to reinforce the unit's main concepts. The Try This! activities are located in the back of the Student Book.

Unit Focus

Take a Chance introduces students to elementary notions of probability. The unit begins with an investigation into fair and unfair situations to motivate discussions of chance. Spinners, number cubes, coins, and other tools are used to explore chance situations. Students estimate probability using everyday language before progressing to percents and part-whole notations. Students discover that repeated trials of an experiment can be used to approximate probabilities. The unit concludes with multiple-event situations that can be represented using tree diagrams.

In working with this unit, students' understanding of probability gradually deepens. First, students should get a basic understanding of some of the situations in which probability arises, and why it might be useful. Next, they should understand that probability ranges on a scale from "sure not to happen" to "sure to happen," from 0% to 100%, and be able to approximate probabilities using this idea. Students should be able to list all possible outcomes of a chance situation. Eventually, students should be able to move to more formal descriptions of probability using fractions, ratios, and percents (although if students are not comfortable with fractions and percents, these more formal representations should not be pushed). Gradual experience with situations requiring more detailed and exact mathematical descriptions will eventually move students into more formal realms of probability.

Mathematical Content

- investigating fair and unfair situations
- using probability and chance in everyday language
- expressing probability as a percent
- expressing probability as a fraction (part-whole language)
- expressing probability as a ratio (one out of ...)
- experimenting with probability
- using tree diagrams
- exploring counting strategies

Prior Knowledge

This unit assumes that students can use simple percents and fractions to describe situations when appropriate and that they can order simple percents and fractions. Students should have an understanding of the mean as introduced in the unit *Picturing Numbers*. In addition, it may be helpful if students have familiarity with coins, number cubes, and spinners.

This unit should be taught after the units *Some of the Parts*, *Measure for Measure*, *Per Sense*, and *Picturing Numbers*.

Facility with adding and subtracting with up to three-digit numbers and with multiplying and dividing two-digit numbers is helpful for this unit. Students should also have some knowledge about reading tables and using a number line to represent and order numbers.

Planning and Preparation

Pacing: 13–15 days

Section	Work Students Do	Pacing*	Materials
A. Fair	■ examine ways to make fair decisions using spinners and number cubes	3 days	■ Letter to the Family (one per student) ■ drawing paper (three sheets per group) ■ number cubes (one per group) ■ scissors, optional (one per group) ■ paste or glue, optional (one bottle per group) ■ large and small paper cups (one per group) ■ bottle cap and chalkboard eraser (one per class) ■ copy of the spinner (one per class) ■ paper clips and brads (one per group) ■ compasses and protractors (one per group) ■ rulers (one per group) ■ coins (one per group) ■ pencils (one per group) ■ tacks (two per group)
B. What's the Chance	■ estimate and describe chances using words, ratios, fractions, or percents	3–4 days	■ Student Activity Sheets 1–3 (one of each per student) ■ drawing paper (two sheets per student) ■ black crayons (one per student)
C. Let the Good Times Roll	■ find chances based on investigations of repeated trials	3 days	■ number cubes (one per group) ■ coins (one per group)
D. Let Me Count the Ways	■ find chances in multiple events ■ use tree diagrams to represent events and to find all possible outcomes	4–5 days	■ Student Activity Sheets 4 and 5 (one of each per student) ■ copies or overhead transparency of the clothes shown on page 28 (one copy per student or one transparency) ■ pennies (two per student) ■ different-colored number cubes (two per pair of students) ■ copies of the chessboard, optional (one per pair of students)

* One day is approximately equivalent to one 45-minute class session.

Preparation

In the *Teacher Resource and Implementation Guide* is an extensive description of the philosophy underlying both the content and the pedagogy of the *Mathematics in Context* curriculum. Suggestions for preparation are also given in the Hints and Comments columns of this Teacher Guide. You may want to consider the following:

- Work through the unit before teaching it. If possible, take on the role of the student and discuss your strategies with other teachers.

- Use the overhead projector for student demonstrations, particularly with overhead transparencies of the student activity sheets and any manipulatives used in the unit.

- Invite students to use drawings and examples to illustrate and clarify their answers.

- Allow students to work at different levels of sophistication. Some students may need concrete materials, while others can work at a more abstract level.

- Provide opportunities and support for students to share their strategies, which often differ. This allows students to take part in class discussions and introduces them to alternative ways to think about the mathematics in the unit.

- In some cases, it may be necessary to read the problems to students or to pair students to facilitate their understanding of the printed materials.

- A list of the materials needed for this unit is in the chart on page xiii.

- Try to follow the recommended pacing chart on page xiii. You can easily spend more time on this unit than the number of class periods indicated. Bear in mind, however, that many of the topics introduced in this unit will be revisited and covered more thoroughly in other *Mathematics in Context* units.

Resources

For Teachers	For Students
Books and Magazines *Mathematics Assessment: Myths, Models, Good Questions, and Practical Suggestions*, edited by Jean Kerr Stenmark (Reston, Virginia: The National Council of Teachers of Mathematics, Inc., 1991) **Videos** *Statistics Strand Teacher Preparation Video*	**Videos** *Mathsphere Videos* • *What Luck* • *What to Expect* • *On Shaky Ground* (available from Encyclopædia Britannica)

Assessment

Planning Assessment

In keeping with the NCTM Assessment Standards, valid assessment should be based on evidence drawn from several sources. (See the full discussion of assessment philosophies in the *Teacher Resource and Implementation Guide*.) An assessment plan for this unit may draw from the following sources:

• Observations—look, listen, and record observable behavior.

• Interactive Responses—in a teacher-facilitated situation, note how students respond, clarify, revise, and extend their thinking.

• Products—look for the quality of thought evident in student projects, test answers, worksheet solutions, or writings.

These categories are not meant to be mutually exclusive. In fact, observation is a key part of assessing interactive responses and also key to understanding the end results of projects and writings.

Ongoing Assessment Opportunities

• **Problems within Sections**
To evaluate ongoing progress, *Mathematics in Context* identifies informal assessment opportunities and the goals that these particular problems assess throughout the Teacher Guide. There are also indications as to what you might expect from your students.

• **Section Summary Questions**
The summary questions at the end of each section are vehicles for informal assessment (see Teacher Guide pages 24, 46, 60, and 86).

End-of-Unit Assessment Opportunities

In the back of this Teacher Guide, there is one assessment that can be given at the end of the unit. It will take approximately two class periods. For a more detailed description of the assessment activity, see the Assessment Overview (Teacher Guide pages 88 and 89).

You may also wish to design your own culminating project or let students create one that will tell you what they consider important in the unit. For more assessment ideas, refer to the charts on pages xvi and xvii.

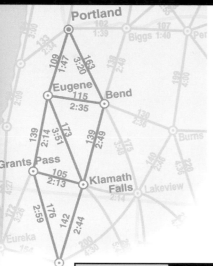

Goals and Assessment

In the *Mathematics in Context* curriculum, unit goals, categorized according to cognitive procedures, relate to the strand goals and the NCTM Curriculum and Evaluation Standards. Additional information about these goals is found in the *Teacher Resource and Implementation Guide.* The *Mathematics in Context* curriculum is designed to help students develop their abilities so that they can perform with understanding in each of the categories listed below. It is important to note that the attainment of goals in one category is not a prerequisite to attaining those in another category. In fact, students should progress simultaneously toward several goals in different categories.

	Goal	Ongoing Assessment Opportunities	End-of-Unit Assessment Opportunities
Conceptual and Procedural Knowledge	**1.** determine whether or not a simple experiment is fair	**Section A** p. 10, #10, #11 p. 14, #17 p. 20, #22	Problems about Chance, p. 107, #3a, #3c
	2. describe chance in everyday language	**Section B** p. 32, #4	Problems about Chance, p. 106, #1 p. 107, #3d
	3. estimate chances in percents, from 0% to 100%	**Section B** p. 32, #5 **Section C** p. 60, #15	Problems about Chance, p. 106, #2 p. 109, #6b
	4. find chances, in percents, fractions, or ratios, for simple situations	**Section B** p. 38, #11 p. 40, #13 p. 42, #15 **Section C** p. 58, #14 **Section D** p. 76, #15	Problems about Chance, p. 106, #2, p. 107, #3d, p. 108, #4b, #4c
	5. list the possible outcomes of simple chance and counting situations	**Section D** p. 70, #10	Problems about Chance, p. 107, #3b

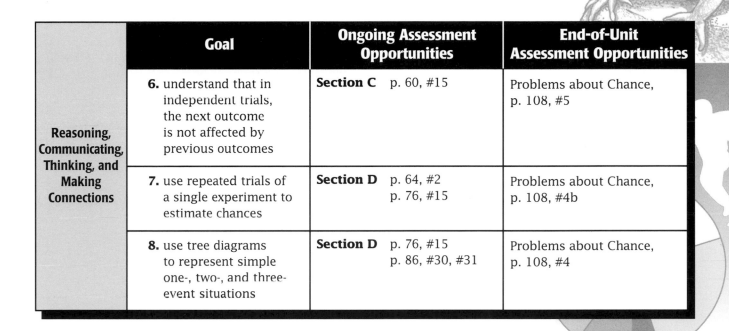

	Goal	Ongoing Assessment Opportunities		End-of-Unit Assessment Opportunities
Reasoning, Communicating, Thinking, and Making Connections	**6.** understand that in independent trials, the next outcome is not affected by previous outcomes	**Section C**	p. 60, #15	Problems about Chance, p. 108, #5
	7. use repeated trials of a single experiment to estimate chances	**Section D**	p. 64, #2 p. 76, #15	Problems about Chance, p. 108, #4b
	8. use tree diagrams to represent simple one-, two-, and three-event situations	**Section D**	p. 76, #15 p. 86, #30, #31	Problems about Chance, p. 108, #4

	Goal	Ongoing Assessment Opportunities		End-of-Unit Assessment Opportunities
Modeling, Nonroutine Problem-Solving, Critically Analyzing, and Generalizing	**9.** understand that variability is inherent in any probability situation	**Section C** **Section D**	p. 52, #7 p. 72, #12c	Problems about Chance, p. 108, #5
	10. model real-life situations involving probability	**Section D**	p. 86, # 30, #31	Problems about Chance, p. 107, #3c p. 109, #6a
	11. develop an understanding of the difference between theoretical and experimental probability	**Section C**	p. 52, #7 p. 60, #15	Problems about Chance, p. 107, #3a p. 108, #5

More about Assessment

Scoring and Analyzing Assessment Responses

Students may respond to assessment questions with various levels of mathematical sophistication and elaboration. Each student's response should be considered for the mathematics that it shows, and not judged on whether or not it includes an expected response. Responses to some of the assessment questions may be viewed as either correct or incorrect, but many answers will need flexible judgment by the teacher. Descriptive judgments related to specific goals and partial credit often provide more helpful feedback than percentage scores.

Openly communicate your expectations to all students, and report achievement and progress for each student relative to those expectations. When scoring students' responses try to think about how they are progressing toward the goals of the unit and the strand.

Student Portfolios

Generally, a portfolio is a collection of student-selected pieces that is representative of a student's work. A portfolio may include evaluative comments by you or by the student. See the *Teacher Resource and Implementation Guide* for more ideas on portfolio focus and use.

A comprehensive discussion about the contents, management, and evaluation of portfolios can be found in *Mathematics Assessment: Myths, Models, Good Questions, and Practical Suggestions,* pp. 35–48.

Student Self-Evaluation

Self-evaluation encourages students to reflect on their progress in learning mathematical concepts, their developing abilities to use mathematics, and their dispositions toward mathematics. The following examples illustrate ways to incorporate student self-evaluations as one component of your assessment plan.

- Ask students to comment, in writing, on each piece they have chosen for their portfolios and on the progress they see in the pieces overall.

- Give a writing assignment entitled "What I Know Now about [a math concept] and What I Think about It." This will give you information about each student's disposition toward mathematics as well as his or her knowledge.

- Interview individuals or small groups to elicit what they have learned, what they think is important, and why.

Suggestions for self-inventories can be found in *Mathematics Assessment: Myths, Models, Good Questions, and Practical Suggestions,* pp. 55–58.

Summary Discussion

Discuss specific lessons and activities in the unit—what the student learned from them and what the activities have in common. This can be done in whole-class discussions, small groups, or in personal interviews.

Connections across the *Mathematics in Context* Curriculum

Take a Chance is the second unit in the statistics strand. The map below shows the complete *Mathematics in Context* curriculum for grade 5/6. This indicates where the unit fits in the statistics strand and in the overall picture.

A detailed description of the units, the strands, and the connections in the *Mathematics in Context* curriculum can be found in the *Teacher Resource and Implementation Guide*.

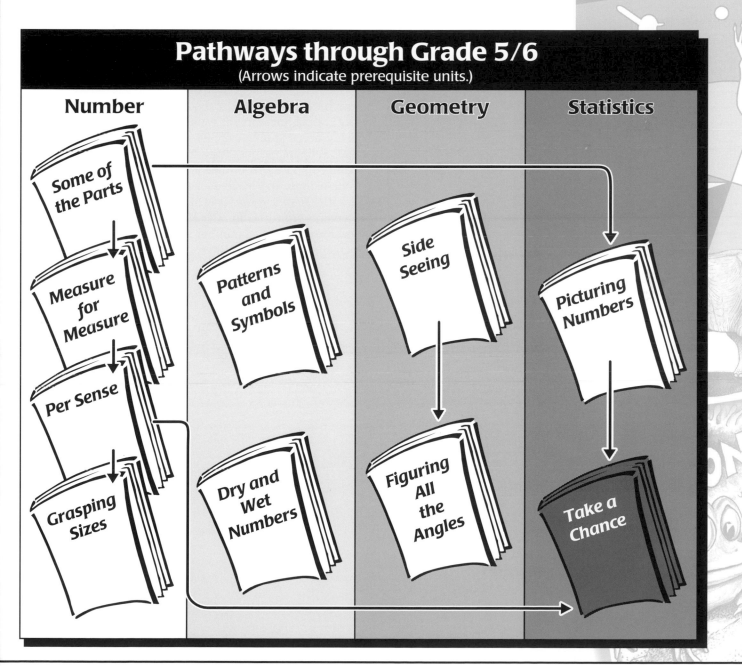

Pathways through Grade 5/6
(Arrows indicate prerequisite units.)

Number	Algebra	Geometry	Statistics
Some of the Parts			
Measure for Measure	Patterns and Symbols	Side Seeing	Picturing Numbers
Per Sense			
Grasping Sizes	Dry and Wet Numbers	Figuring All the Angles	Take a Chance

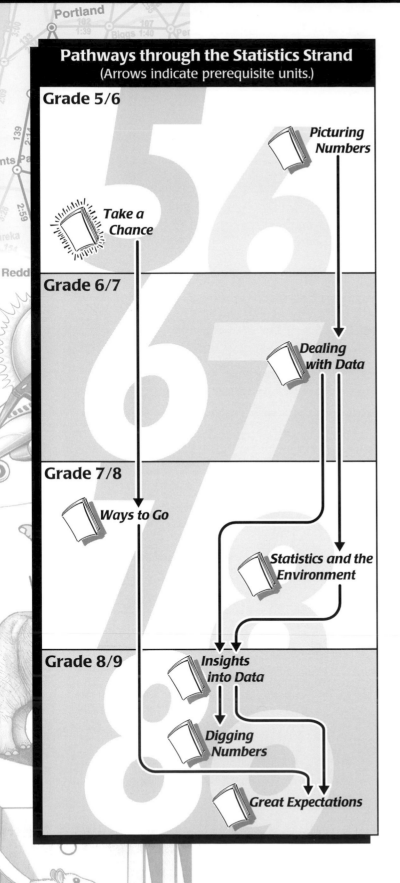

Pathways through the Statistics Strand
(Arrows indicate prerequisite units.)

Grade 5/6

Picturing Numbers

Take a Chance

Grade 6/7

Dealing with Data

Grade 7/8

Ways to Go

Statistics and the Environment

Grade 8/9

Insights into Data

Digging Numbers

Great Expectations

Connections within the Statistics Strand

On the left is a map of the statistics strand; this unit, *Take a Chance*, is highlighted.

The major themes in the statistics strand are dealing with data, developing an understanding of chance and probability, using probability in situations connected to statistics, and developing critical thinking skills. *Take a Chance*, the second unit in the statistics strand, provides students with initial experiences in quantifying chance and making fair decisions.

The conceptual themes in the statistics strand are data collection, data visualization, numerical characteristics, and reflections and conclusions. *Picturing Numbers* introduces these concepts.

In *Picturing Numbers*, students are introduced to graphic representations of simple data: pie charts, bar graphs, pictographs, number line graphs, and line graphs. Students create, investigate, interpret, and compare graphs. They collect and then graph data, compare sets of data and graphs, draw conclusions, and make decisions based on the graphs. *Picturing Numbers* also contains elements from the number strand—students use measurements, number sense, fractions, and percents to make and interpret different types of graphs.

The Statistics Strand

Grade 5/6

Picturing Numbers
Creating and interpreting different kinds of graphs (bar and line graphs, pie charts, and pictographs), describing data in tabular and graphical forms, and using data as an aid to argument.

Take a Chance
Exploring the meanings of the terms *fair* and *chance*; computing chances in situations with few outcomes; using repeated trials to estimate chances; and using tree diagrams to represent two-or three-event situations.

Grade 7/8

Ways to Go
Reading and interpreting different kinds of maps, drawing graphs and networks, representing tables as graphs (and vice versa), solving problems with the help of connectivity graphs and tables, and solving simple combinatoric problems. (*Ways to Go* is also in the geometry strand.)

Statistics and the Environment
Collecting data; summarizing data with one-number summaries; reading and interpreting different kinds of maps, graphs, and tables; and using graphs and data as aids to argument

Grade 6/7

Dealing with Data
Creating and interpreting scatter plots, box plots, stem-and-leaf diagrams, histograms, and number line plots; describing data in tabular and graphical forms; using one-number summaries (mean, median, mode, range, quartiles); and using data as an aid to argument.

Grade 8/9

Insights into Data
Analyzing information presented in a graph; gathering data using sampling techniques; recognizing bias in sample surveys; examining misleading representations of data; identifying weak, strong, positive, and negative correlations and no correlation; using lines to summarize the data in a scatter plot; making and interpreting scatter plots and box plots.

Digging Numbers
Using the properties of height, diameter, and radius to determine whether or not various irregular shapes are similar; predicting length using graphs and formulas; exploring the relationship between three-dimensional shapes and drawings of them; and using length-to-width ratios to classify various objects. (*Digging Numbers* is also in the geometry strand.)

Great Expectations
Recognizing the variability in taking random samples and the role of chance in world events; understanding the effect of sample size on the variability in samples; finding the probability of independent and dependent events; using expected values to make decisions; using simulations to investigate probability; and counting all the possible arrangements of a set, with and without repetition

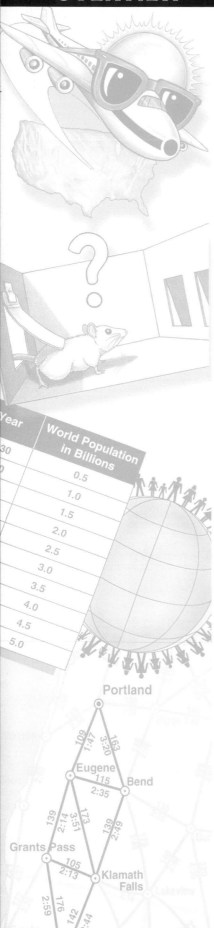

Connections with Other *Mathematics in Context* Units

Chances are often represented as fractions, decimals, percents, ratios, or proportions. This explains the strong connections that this unit has to units in the number strand. Visual models are often useful to represent chances. In this sense, one may make a connection with units in the geometry strand. For example, to make fair spinners, students need to have some notion of angles as developed in the unit *Figuring All the Angles*. The tree diagrams and the tables used to record the events in *Take a Chance* are organizational models and tools that also appear in the unit *Digging Numbers*.

The following mathematical topics that are included in the unit *Take a Chance* are introduced or further developed in other *Mathematics in Context* units.

Prerequisite Topics

Topic	Unit	Grade
fractions	*Some of the Parts**	5/6
percents	*Per Sense**	5/6
decimals	*Measure for Measure**	5/6
pie charts, data collection and data representation	*Picturing Numbers*	5/6
angles and turns	*Figuring All the Angles****	5/6
number line, ratio table	*Number Tools*	5/6

Topics Revisited in Other Units

Topic	Unit	Grade
counting strategies	*Ways to Go*	7/8
probability	*Ways to Go*	7/8
	Great Expectations	8/9
tree diagrams	*Ways to Go*	7/8
	Great Expectations	8/9
ratio and proportion	*Grasping Sizes**	5/6
	*Looking at an Angle****	7/8
fractions	*Fraction Times**	6/7
	*Looking at an Angle****	7/8
percents	*Fraction Times**	6/7
	*More or Less**	6/7
angles and shapes	*Triangles and Beyond****	7/8
	*Looking at an Angle****	7/8
tables and classification	*Digging Numbers***	8/9

   * These units in the number strand also help students make connections to ideas about statistics and probability.

  ** These units in the algebra strand also help students make connections to ideas about statistics and probability.

 *** These units in the geometry strand also help students make connections to ideas from statistics and probability.

Student Materials and Teaching Notes

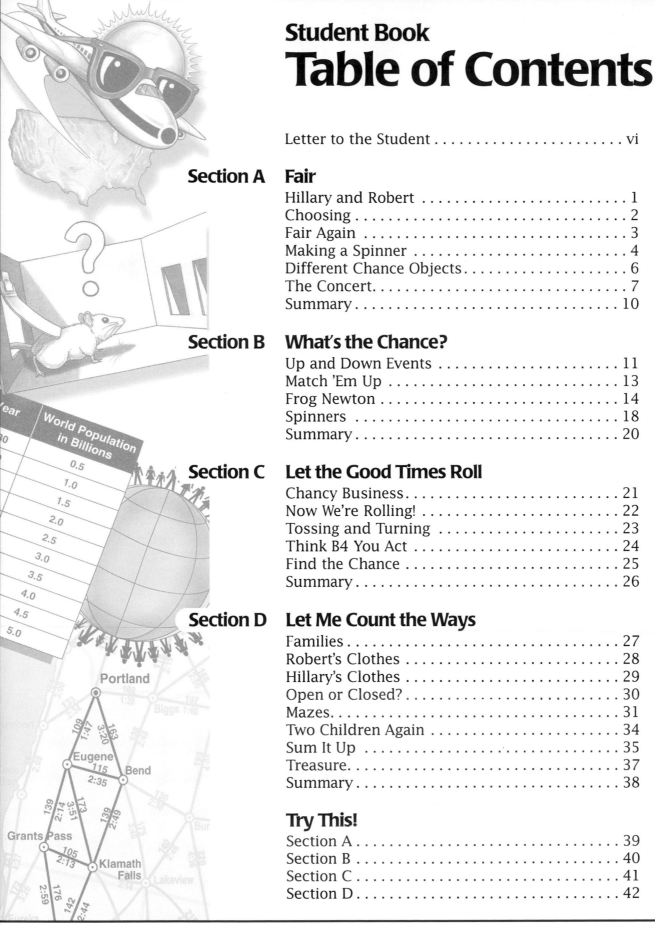

Student Book
Table of Contents

Dear Student,

You are about to begin the study of the *Mathematics in Context* unit *Take a Chance*. Think about the following words and what they mean to you: *fair, sure, uncertain, not likely, impossible*. In this unit, you will see how these words are used in mathematics.

You will toss coins and number cubes and record the outcomes. Do you think you can predict how many times a head will come up if you toss a coin a certain number of times? Is the chance of getting a head greater than the chance of getting a tail? As you investigate these ideas, you are beginning the study of probability.

When several different things can happen, you will learn how to count all of the possibilities in a "smart" way. Keep alert during the next few weeks for statements that you may read or hear, such as "The chance of rain is 50%." You might even keep a record of such statements and bring them to share with the class.

We hope you enjoy learning about chance!

Sincerely,

The Mathematics in Context Development Team

Work Students Do

Students learn how to make fair decisions using coins, number cubes, spinners, and other objects as tools. Students decide who will be first to play a computer game, who will do a class report on dinosaurs from one of three geographical regions, and who will get to go to a rock concert. Finally, students evaluate the fairness of decisions made in real-life situations involving sports, school activities, elections, and television programs.

Goals

Students will:

- determine whether or not a simple experiment is fair;
- describe chance in everyday language;*
- list the possible outcomes of simple chance and counting situations;*
- use repeated trials of a single experiment to estimate chances;*
- understand that variability is inherent in any probability situation;*
- model real-life situations involving probability.*

  * These goals are introduced in this section and assessed in other sections in the unit.

Pacing

- approximately three 45–minute class sessions

Vocabulary

- chance
- fair

About the Mathematics

Students beginning the study of probability should already have some intuitive understanding of chance concepts as expressed in everyday language such as *it will probably happen* or *it probably won't happen*. In this section, determining what is fair and what is not fair motivates discussions of chance. Students should get a sense of the number of possible outcomes using different objects such as spinners and number cubes. Although at this stage, students are not asked to use percents or fractions to describe chance; students should begin to see that the number of outcomes is important to consider when deciding whether or not something is fair.

Materials

- Letter to the Family, page 100 of the Teacher Guide
- drawing paper, pages 11, 13, and 15 of the Teacher Guide (three sheets per group of students)
- number cubes, pages 11 and 19 of the Teacher Guide (one per group of students)
- scissors, page 11 of the Teacher Guide, optional (one pair per group of students)
- paste or glue, page 11 of the Teacher Guide, optional (one bottle per group of students)
- large paper cup, page 17 of the Teacher Guide (one per class)
- small paper cup, page 17 of the Teacher Guide (one per class)
- bottle cap, page 17 of the Teacher Guide (one per class)
- chalkboard eraser, page 17 of the Teacher Guide (one per class)
- copy of the spinner, page 17 of the Teacher Guide (one per class)
- paper clips, pages 13 and 15 of the Teacher Guide (one per group of students)
- brads, pages 13 and 15 of the Teacher Guide (one per group of students)
- compasses, pages 13 and 15 of the Teacher Guide (one per group of students)
- protractors, pages 13 and 15 of the Teacher Guide (one per group of students)
- rulers, pages 13 and 15 of the Teacher Guide (one per group of students)
- coins, page 19 of the Teacher Guide (one per group of students)
- six-sided pencils, page 25 of the Teacher Guide (one per group of students)
- tacks, page 25 of the Teacher Guide (two per group of students)

Planning Instruction

This section is an introduction to chance. It builds on students' informal knowledge of chance and probability. You may have a short class discussion at the start of the unit to introduce the topic of chance. In this section students perform a number of experiments. You may want to have all the materials needed for these experiments available before you start the section. If time is a factor, you may decide to have the students do some of the experiments at home after they have been introduced in class.

Have students work on problem 1 as a class. Pair students for problems 13 and 14. Students may work on the remaining problems in pairs or in small groups.

Problem 24 is optional and can be skipped if time is a concern.

Homework

Problem 21 (page 18 of the Teacher Guide) can be assigned as homework. Also, the Extensions (pages 11 and 23 of the Teacher Guide) and the Writing Opportunity (page 25 of the Teacher Guide) can be assigned as homework. After students complete Section A, you may assign appropriate activities from the Try This! section, located on pages 39–42 of the *Take a Chance* Student Book. The Try This! activities reinforce the key math concepts introduced in this section.

Planning Assessment

- Problems 10, 11, 17, and 22 assess students' ability to determine whether or not a simple experiment is fair.

A. FAIR

This unit follows Hillary and Robert, students at Eagle School in Maine, as they experiment with using **chance** to make decisions.

You probably already know some things about chance.

1. What do you think of when someone says the word *chance*?

1. Answers will vary. Sample student responses:

   - I think of the chances of rain. For example, there is a 40 percent chance that it will rain tomorrow.

   - I think of the chances of our team winning the next game. For example, there is a 50-50 chance that our team will win the next game.

   - I think of the chances of winning the lottery. For example, the chances of winning the lottery are very small.

Overview Students describe what the word *chance* means to them.

About the Mathematics Problem **1** will help you evaluate students' knowledge of probability. Students should have some intuitive understanding of chance concepts as expressed in everyday language such as *it will probably happen* or *it probably won't happen.*

Planning You may have a short class discussion about problem **1** to introduce this first section. Do problem **1** as a class.

Comments about the Problems

1. Students' personal experiences will vary. Some students may describe chance in general terms (a small chance, a sure thing), while others may use more exact language (a 10% chance). A 100% chance means that there is a certainty that something will happen. At this point, it does not matter if students do not talk about chance using fractions or percents. These topics will be discussed later in the unit.

Choosing

Hillary and Robert both want to play *Super Math Wiz III*, a computer game, during lunch. The game is installed on one computer in the classroom. Since only one person can play at a time, Hillary and Robert have to decide who will play the game first.

2. How would you solve this problem in a ***fair*** way?

3. What do you think it means to be *fair*?

There are many situations in which you have to find a fair way to make a decision.

Robert says to Hillary, "If you throw a 6 with this number cube, you can play; otherwise, I'll play."

4. Do you think this is fair? Why or why not?

5. Can you come up with a better way to decide?

Hillary and Robert finally decide to use a spinner like the one on the left. They decide to spin once. If the arrow points to black, Hillary will go first. If the arrow points to white, Robert will go first.

6. Is this a fair method? Why or why not?

2. Answers will vary. Sample student responses:

- by flipping a coin,

- by rolling a number cube (each person gets three numbers), or

- by spinning a spinner (with the sections of the spinner equally divided between the two players).

3. Answers will vary. Sample student responses:

- Fair means everyone has an equal chance of being chosen or of winning.

- Fair means that each number on a number cube has an equal chance of coming up.

- Fair means that when a student is chosen to run an errand, the teacher does not play favorites.

- Fair means taking turns so no one is left out.

- Fair means equal.

4. This is not fair. A 6 will come up much less often than the other five numbers.

5. Answers will vary. Sample student responses:

- Divide the six numbers on the number cube equally between Hillary and Robert. For example, assign the numbers 1, 2, and 3 to Hillary and the numbers 4, 5, and 6 to Robert.

- Toss a coin. If it lands on heads, Hillary goes first, and if it lands on tails, Robert goes first.

- Divide sections of a spinner equally between the two players. Then spin the spinner to see who will play first.

6. The method is fair. Explanations will vary. Sample explanation:

The chance of spinning black is equal to the chance of spinning white because the spinner is divided into two equal sections.

Overview Students solve a familiar problem: how to decide in a fair way who will play first in a game. They also discuss how to make decisions in a fair way using a number cube or a spinner.

About the Mathematics Students should be allowed to explore what the word *fair* means, and what it might mean in the context of chance. Determining what is fair and what is not fair motivates the discussion of chance. This will lead to the study of the number of possible outcomes using different objects, such as number cubes and spinners.

For a decision-making method to be fair, not only must the possible outcomes be divided equally between the two students, but each outcome must be as likely as any other. For example, look at these two spinners:

Each spinner is divided into two parts, but the spinner on the right is not fair because landing on white is more likely than landing on black. A similar situation arises in problem **4.** The method used to determine who will play the game first is unfair because the outcomes are not divided equally between the two players.

Planning Students may work in pairs or in small groups on problems **2–6.** Discuss problems **2** and **3** with the whole class.

Comments about the Problems

- **2–3.** Some students may suggest methods that are unfair. You may want to record the methods students propose on the chalkboard and discuss whether or not each method is fair. You can also discuss the chance involved in each method.

- **4.** Although some students may use phrases such as *one out of six,* do not use fractions at this point. Fractions will be introduced in Section B.

By now, everyone in Hillary's and Robert's class has heard about the computer game and wants to play. Hillary says, "Okay, okay! Let's put all of our names in a hat, and the person who is picked gets to play."

7. Is this fair? Why or why not?

A method for choosing is *fair* if it gives everyone the same chance of being chosen.

8. Think of two other situations in which it is important to be fair.

Fair Again

Hillary and Robert both want to play the computer game again the next day. They decide to toss a coin to see who will play first. Since there is an equal chance (sometimes called a 50-50 chance) of getting either heads or tails, this is a fair method.

You can draw a diagram with branches like a tree to show the two possibilities. The path you take on the tree shows the side of the coin that came up.

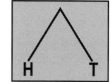

9. a. What do the H and T stand for?

   b. Robert says, "You know, the diagram shows that there's a 50-50 chance of getting a head or a tail." Explain what Robert means.

Look at a number cube.

10. a. How many different numbers can you roll on a number cube?

   b. Draw a picture to show the different possibilities.

11. How could Robert and Hillary use a number cube to decide in a fair way who will play the game first?

12. Hillary and Robert have a black-and-white cube. Hillary wins if it comes up white, and Robert wins if it comes up black. How can you tell if the cube has been colored in a fair way?

7. Answers will vary. Sample response:

This is a fair method only if all the names are written on pieces of paper that are the same size and if everyone's name is put into the hat only once (or an equal number of times).

8. Answers will vary. Students' two responses may include the following:

- when dividing food
- when choosing teams
- when judging a contest
- when grading students' papers

9 a. H stands for heads; T for tails.

   b. Fifty-fifty means that heads and tails have equal chances of coming up when a coin is tossed.

10. a. six

   b. Pictures will vary. Sample student drawings:

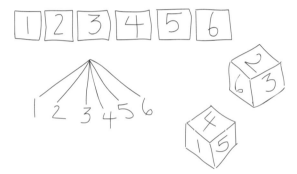

11. Answers will vary. Sample response:

Three numbers may be assigned to each person. For example, Hillary could get 1, 2, and 3, while Robert gets 4, 5, and 6. They roll the number cube. The person whose number comes up goes first.

12. There must be three white and three black sides. There are many possibilities. The sides with the numbers 1, 2, and 3 could be colored black. The sides with the numbers 4, 5, and 6 could be colored white.

Materials number cubes (one per group of students); drawing paper (two sheets per group of students); scissors, optional (one pair per group of students); paste or glue, optional (one bottle per group)

Overview Students explore tossing a coin, rolling a number cube, and taking names out of a hat to determine which method could fairly decide who should play a computer game first. They also draw diagrams to show the possible outcomes of tossing a coin and rolling a number cube.

About the Mathematics A simple tree diagram is used to show the two possible outcomes of tossing a coin. It is not necessary to discuss this model extensively at this time. Students investigate tree diagrams in greater depth in Section D.

Planning Students may continue to work in pairs or in small groups on problems **7–12.** Problems **10** and **11** may be used as assessments. After students finish problem **12,** they should have a good understanding of the concept of *fair* and be able to explain what is fair in different real-life situations. If some students are having difficulty with the concept, discuss problems **4, 5,** and **10** with the whole class.

Comments about the Problems

   9. Students' answers will show their informal knowledge about percents and chance. *Fifty-fifty* refers to the percent chance of a coin's landing on heads or on tails. Students may remember this situation from the grade 5/6 unit *Per Sense.*

10–11. Informal Assessment These problems can be used to assess students' ability to determine whether or not a simple experiment is fair.

   10. The problem covers the same topic as problem **9,** but now there are six possible outcomes instead of two.

   11. Some students may say that problem **11** is the same as problem **5.**

Extension Ask students to design a cutout (or net) for a number cube at home. They can then cut it out and paste it together in class the next day. You might also challenge students with the following question: *On what side of the cube should each number be written so that opposite sides of the cube always add up to seven?* [1 on the opposite side of 6; 2 on the opposite side of 5; and 3 on the opposite side of 4] These number cubes can be used for many activities, especially in Section C.

Activity

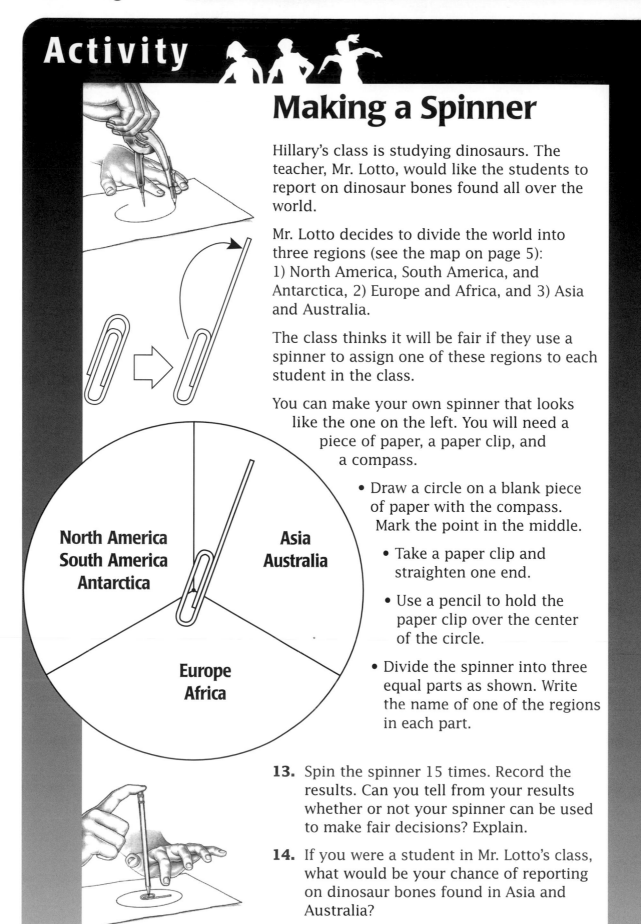

Making a Spinner

Hillary's class is studying dinosaurs. The teacher, Mr. Lotto, would like the students to report on dinosaur bones found all over the world.

Mr. Lotto decides to divide the world into three regions (see the map on page 5): 1) North America, South America, and Antarctica, 2) Europe and Africa, and 3) Asia and Australia.

The class thinks it will be fair if they use a spinner to assign one of these regions to each student in the class.

You can make your own spinner that looks like the one on the left. You will need a piece of paper, a paper clip, and a compass.

- Draw a circle on a blank piece of paper with the compass. Mark the point in the middle.

  - Take a paper clip and straighten one end.

    - Use a pencil to hold the paper clip over the center of the circle.

  - Divide the spinner into three equal parts as shown. Write the name of one of the regions in each part.

North America South America Antarctica

Asia Australia

Europe Africa

13. Spin the spinner 15 times. Record the results. Can you tell from your results whether or not your spinner can be used to make fair decisions? Explain.

14. If you were a student in Mr. Lotto's class, what would be your chance of reporting on dinosaur bones found in Asia and Australia?

13. Results and answers will vary. Sample responses:

Outcome	Number of Times It Came Up
North and South America and Antarctica	|||
Asia and Australia	✝HT ||
Europe and Africa	✝HT

You cannot tell from the results that the spinner is fair. But the spinner is divided into three equal parts, and each part has an equal chance of coming up. So, the spinner is fair.

14. Some students may say that the chance of landing on Asia and Australia is the same as the chance of landing on each of the other two areas. Other students may say that the chance of landing on Asia or Australia is one out of three, or one-third.

Materials drawing paper, paper clips, brads, compasses, protractors, and rulers (one of each per group of students)

Overview Students construct spinners and experiment with them to determine whether or not they could be used to fairly assign report topics in class.

About the Mathematics In this activity, students are informally introduced to the concepts of angles and turns. These concepts are studied more extensively in the grade 5/6 unit *Figuring All the Angles*.

Planning Students may construct the spinners and do problems **13** and **14** in pairs. Students may want to use a brad, rather than a pencil, to hold the paper clip in place on the spinner. They must realize that in order for the spinner to have equal sections, each angle must measure 120°. Discuss the results of problem **13** with the whole class. Encourage students to pay close attention to the outcomes and distribution of spins.

Comments about the Problems

13. Students should be allowed to spin more often if they like. You may combine students' results, although it is time consuming and will be done again later in the unit. Encourage students to begin each spin in a different place and with a different amount of force. Otherwise, they may end up with the same result after each spin. Suggest that they make a tally or a chart to record their results.

14. Some students may recognize the pie pieces in the picture of the spinner and use fractions to represent chance. Using fractions to represent chance is discussed in Section B.

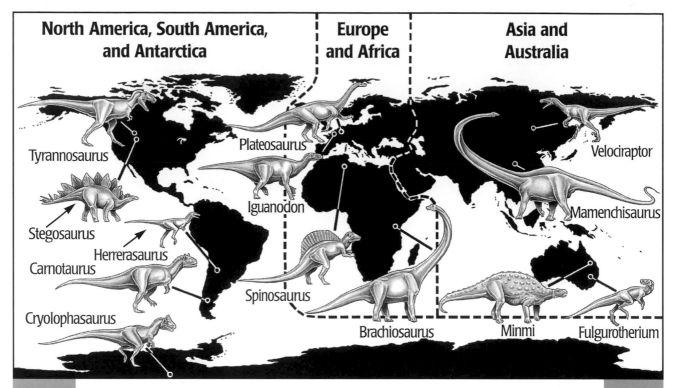

North America, South America, and Antarctica

Europe and Africa

Asia and Australia

Tyrannosaurus

Plateosaurus

Velociraptor

Stegosaurus

Iguanodon

Mamenchisaurus

Herrerasaurus

Carnotaurus

Spinosaurus

Cryolophasaurus

Brachiosaurus

Minmi

Fulgurotherium

△ Shanna wonders whether or not a spinner made out of a triangle can be used to make fair decisions.

15. a. Draw a triangle on paper. Can you make the triangle into a fair spinner?

   b. How can you tell whether it is fair or not?

   c. Can any triangle be made into a fair spinner? Support your answer with some examples.

16. a. Jonathan wonders if he can use a number cube to choose regions in a fair way. Can he? If so, how?

   b. Kara wonders if she can use a coin. Can she?

17. How can you tell whether or not a particular method will be fair? Explain your answer.

18. Since many bones were recently found in Europe and Africa, Mr. Lotto thinks there should be more students reporting on this region than on the other regions.

   a. How would you make a spinner so that the region "Europe and Africa" is picked more often?

   b. Would this spinner be fair?

15. a. Yes. Sample student drawing:

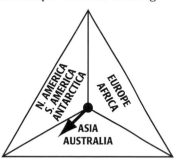

b. It is a fair spinner if the interior angles for the regions are equal. Note that the areas do not have to be equal, only the angles do. Some students may understand this point. Sample student drawing:

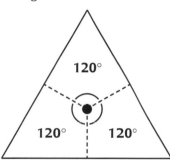

c. Yes. The only requirement is that the spinner be divided into equal angles. Allow students to experiment.

16. a. A number cube will work if each region is assigned two numbers on the cube.

b. A coin has only two outcomes, so it will not work.

17. In this instance, a fair method has a total number of outcomes that is divisible by three, with each outcome having the same chance of occurring.

18. a. You would make the central angle that represents Europe and Africa larger than the central angles for the other regions. Another possibility is to divide the spinner into more than three equal sections and give Europe and Africa more sections than the other regions.

b. No, this spinner would not be fair for all regions.

Materials drawing paper (one sheet per group of students); paper clips, brads, compasses, protractors, and rulers (one of each per group of students)

Overview Students construct a spinner in the shape of a triangle and determine whether or not it could be used to fairly assign class report topics. Then students consider using a number cube or a coin to assign report topics.

Planning Students may continue to work in pairs or in small groups on problems **15–18.** You may use problem **17** for assessment. You may want to discuss problem **15** with the whole class.

Comments about the Problems

15. Note that the areas of the triangle sections do not have to be equal, but the angles around the central point do. To emphasize this point, look at a similar situation. Ask students these questions: *Have you ever seen a watch in a shape other than a circle? How could you design that watch face so it could be used to tell time?* Here are two possible answers:

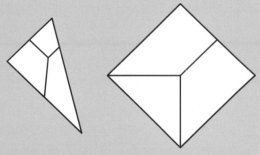

16. b. Students may try to come up with elaborate schemes for using a coin, but theses schemes will probably not offer equal chances. One sophisticated method involves tossing three coins, but it is too soon to discuss this. A similar situation is presented in Section D, problem **10.**

17. Informal Assessment This problem assesses students' ability to determine whether or not a simple experiment is fair.

18. Again, the sizes of the central angles, not the areas of the sections, are important.

Activity

Different Chance Objects

Hillary wonders whether or not objects other than spinners and number cubes can be used to make fair decisions. When objects are not shaped as regularly as coins, number cubes, or spinners, it can be hard to tell. One way to find out is to flip or spin the object over and over again to see what happens each time.

Your teacher will divide the class into groups. Each group will get one of the items listed below:

- a large paper cup
- a small paper cup
- a chalkboard eraser
- a bottle cap
- the spinner on the left

Your job is to find out whether or not you can use your item in any way to make a fair decision.

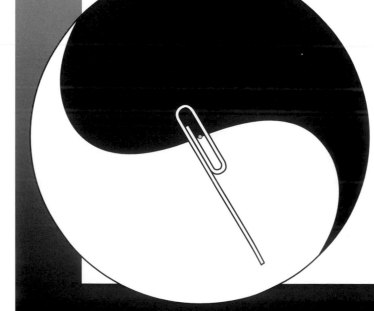

19. Throw or spin your item 30 times. Make a table of your results. When you are done, decide whether or not you can use your item to make a fair decision. Report your results to the class.

Note: Keep these results, because you will use them again later in the unit.

19. The cups and eraser cannot be used to make a fair decision, while the spinner can be used in such a way. The bottle cap may or may not be used to make a fair decision, depending on the individual bottle cap.

Sample student tables:

Cup:
3 trials

Side	26	27	28
Top	3	3	1
Bottom	1	0	1

Eraser:
4 trials

Side 1	5	3	1	2
Side 2	4	0	6	2
Top	10	15	9	14
Bottom	11	12	14	12

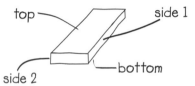

The eraser never got on the short sides.

Materials large and small paper cups, a bottle cap, a chalkboard eraser, and a copy of the spinner on page 6 of the Student Book (one per classroom)

Overview Students experiment with irregularly shaped objects and determine whether or not these objects can be used to make fair decisions.

About the Mathematics Regularly-shaped objects offer equal chances of landing on any given side, and these chances can be calculated in advance. The objects students use in this activity have irregular shapes; therefore, the chances of the possible outcomes must be determined experimentally. This problem may be used to introduce the idea that variability occurs in any probability situation. If more than one group uses the same item, discuss the similarities and differences in the results. For example, when flipping a bottle cap 50 times, two groups got these results:

	H	T
Group 1	26	24
Group 2	23	27

You may ask students if this is what they would have expected, and to explain why or why not.

Planning Prepare the materials for this activity in advance. You can have each group experiment with more than one item. Students may continue to work on problem **19** in pairs or in small groups. Ask students to save their work because this activity will be used again in Section C, on page 25 of the Student Book.

Comments about the Problems

19. You may want to discuss the possible outcomes of each object with the group using it. The number of outcomes for each object depends on whether or not you accept some of the more unlikely flips as possibilities. If all outcomes are acceptable, then the cap and spinner have two outcomes each, the cup has three outcomes, and the eraser has six outcomes (although the possibility of the eraser landing on an end is very unlikely).

The Concert

Next week Compass Rose, a rock band that Hillary likes, is coming to play in Eagle. Hillary's mother got four tickets to the concert. She will take Hillary and two of her friends.

Unfortunately, Hillary has three friends she wants to bring and has to find a fair way to decide who will go with her.

20. Find a fair way to decide which two friends will go with Hillary. You may use coins, number cubes, spinners, or anything else you think may be fair.

Oh no! Another of Hillary's friends wants to go too!

21. Come up with a fair way to decide which two of the four will go with Hillary now.

20. Answers will vary. Sample responses:

- Using a number cube, assign two different numbers to each friend. Roll the number cube 30 times and record the results. The two friends whose numbers come up most often get to go with Hillary.

- Divide a spinner into three equal sections. Spin the spinner to choose friends. On the second spin, if the spinner lands on the name of the friend already chosen, spin the spinner again until a second friend is chosen.

- Write the friends' names on equal-size pieces of paper. Put their names in a hat and choose one.

21. Answers will vary. The easiest method is to write the four friends' names on pieces of paper and draw their names from a hat. A spinner divided into four equal sections could also be used.

Materials coins, number cubes, or spinners (one of each per group of students)

Overview Students decide which method they will use to choose who will go to a rock concert when a limited number of tickets are available. They can use coins, spinners, number cubes, or any other object.

About the Mathematics In these problems, students focus on the number of possible outcomes. When there are three choices, the method for choosing must have three (or a multiple of three) outcomes. With four choices, there must be four (or a multiple of four) outcomes.

Planning After students complete problem **21,** they should be able to explain how to use spinners, coins, or number cubes to make fair decisions. If some students are having difficulty, discuss this problem in small groups or with the whole class. Students may work on problems **20** and **21** in pairs or in small groups. You may assign problem **21** for homework.

Comments about the Problems

20. A spinner with three equal sections can be used to solve this problem. However, if the first person's number or name reoccurs, repeated trials may have to be conducted to select the second person. A more direct approach using the spinner is to choose the person who will not go.

A coin is difficult to use in this problem. Some students may try to solve the problem by tossing a coin, assigning heads for friend A and tails for friends B and C. If tails comes up on the first toss, friends B and C get to go with Hillary. If heads comes up on the first toss, friend A gets to go and a second toss is required to decide between friends B and C. With this method, friend A has a 50% chance of being picked, while friends B and C have only a 25% chance each.

21. Homework This problem may be assigned as homework. It is possible to solve this problem by tossing two coins. It is not very likely that students will come up with this method, because some students may fail to realize that there are four possible outcomes: HH, HT, TH, and TT. Many students might say there are three possible outcomes: TT, HH, and either HT or TH.

22. Give your opinion about the fairness of each of the following situations:

a. Two soccer teams toss a coin before a game to see which team gets to choose a goal.

b. In Mr. Ryan's class, there are 10 boys and 15 girls. To decide who will be hall monitors each day, Mr. Ryan draws the name of one girl from a box holding all of the girls' names and the name of one boy from a box holding all of the boys' names.

22. a. This is a fair situation. Sample explanations:

- There is an equal chance of getting either heads or tails on the coin toss.

- There is a 50-50 chance of heads or tails coming up.

- The diagram shows that their chances are equal.

b. This is not a fair situation. In this method, boys have a higher chance of being chosen because there are fewer boys in the class.

Overview Students give their opinions about the fairness of two real-life situations.

Planning Students may continue to work in pairs or in small groups. You may want to use problem **22** for individual or for group assessment. Be sure to discuss problems **a** and **b** with the whole class.

Comments about the Problems

22. Informal Assessment This problem assesses students' ability to determine whether or not a simple experiment is fair.

22. a. Part **a** presents the same situation as problem **9.** Remind students to write down their explanations.

b. In part **b,** Mr. Ryan probably wants to pick one boy and one girl. Because there are fewer boys than girls, a boy has a greater chance of being chosen. Some students may respond that a boy has a 1 out of 10 chance of being chosen and a girl has a 1 out of 15 chance of being chosen. If students do not realize that the numbers of boys and girls influence the chances of being selected, you can suggest the following: *What would happen if there were one boy and 15 girls in that class?* [That boy would be chosen.] *What numbers of boys and girls would be needed to make this method fair?* [10 boys and 10 girls, or 15 boys and 15 girls, or any equal number.]

c. Only 50 students can go on a field trip to the zoo because there is only one bus. The principal decides to allow the first 50 who sign up before school in the morning to go on the trip.

d. In the United States, all people 18 years old or older are eligible to vote for a presidential candidate.

22. c. This is not fair because some students may not be able to get to school early. Everyone does not have an equal chance.

d. Answers will vary. This is not a question of mathematical fairness, but one of opinion. Again, this problem may require discussion about whether or not people under 18 should be allowed to vote.

Overview Students give their opinions about the fairness of two more real-life scenarios.

Planning You may want to continue to use this problem for individual assessment or for group assessment. These two problems differ from parts **a** and **b** because "fair" is presented here in nonmathematical contexts. This may require some discussion before students begin.

Comments about the Problems

22. d. Any person 18 years or older can vote for any presidential candidate. So, the elections are mathematically fair. Of course, many other factors play a role.

Extension Have students play the following game. Then challenge them to find out why the game is unfair. The game is played with two players. Player 1 starts by saying the number "one." Player 2 must increase the number by one or two. Player 1 then increases the number by one or two. The two players take turns. The winner is the first person to say "ten." [The game is unfair because Player 1 can always win by saying the numbers one, four, seven, and ten.]

Summary

There are many situations in daily life that involve chance or in which you must make fair decisions. *Fair* means that every possibility has the same chance of occurring.

In order to make fair decisions, you should use a *fair* method. Typical things that can help you make fair decisions are coins, number cubes, and spinners. There are many other objects that can help you make fair decisions.

Summary Questions

23. a. Can you toss a pencil to make a fair decision? How might you find out?

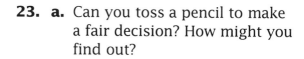

   b. Can you toss a thumbtack to make a fair decision? How might you find out?

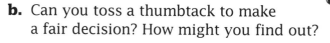

A quiz show on television features two teams from local schools that compete every week. A participating school may send 40 students to sit in the audience. The principal of Eagle School has decided that each of the eight classes in the school should select—by a drawing—five students who will go to the studio.

On the left is a list of the eight classes at Eagle School.

The principal says that her method is fair.

Class	Number of Students	Number Selected
Ms. Johnson	25	5
Mr. Geist	27	5
Ms. Lanie	32	5
Ms. McGill	31	5
Mr. Ford	24	5
Ms. Durden	25	5
Mr. Shore	29	5
Mr. Lane	32	5

24. a. If you were a student at Eagle School, which class would you want to be in?

   b. Is the principal's method fair?

   c. Hillary and Robert have decided to design a different method that will be fair for the principal to use. What method do you think they might come up with?

23. a. Yes, a pencil can be used for fair decisions if:

1) the pencil has regularly shaped sides,

2) the chance of landing on the eraser is very small, or

3) you number the sides.

You might find out by flipping a pencil 100 times, or just by looking at its sides.

b. No. A thumbtack is too irregularly shaped to be predictable. You would have to toss it many times to determine whether or not it can be used to make a fair decision.

24. a. Mr. Ford's class. A student in the class with the fewest total students has the best chance of being chosen.

b. No, because students in different classes do not have the same chance of being chosen.

c. Answers will vary. One fair method would be to put the names of all of the students in the school in one hat and pick 40 at random.

Materials tacks (two per group of students), six-sided pencils (one per group of students)

Overview Students read the Summary, which reviews the main concepts covered in this section. They then evaluate the fairness of different methods used to make decisions.

About the Mathematics In Section B, students will learn to describe the chance that an event will occur using numbers (fractions, percents, and ratios). On this page, allow the students to use more qualitative descriptions, such as *a higher chance, same chance,* or *small chance.*

Planning Have students read the Summary and work in pairs or in small groups on problems **23** and **24.** If time is a concern, you may skip problem **24.** After students complete Section A, you may assign appropriate activities from the Try This! section, located on pages 39–42 of the *Take a Chance* Student Book, for homework.

Comments about the Problems

23. If you do not want students to use tacks, demonstrate flipping a tack on an overhead projector. The shadow of the tack can easily be seen.

24. This problem is similar to problem **22b.**

Writing Opportunity You may ask students to write their answers to problem **22** in their journals.

Work Students Do

Students begin by predicting whether or not given events will take place. They then estimate chances for various events and label a "chance" ladder with these events. Using the context of a frog's hopping on a black-and-white tile floor, students describe the chance that the frog will land on a black tile using percent, a fraction, and a ratio. Lastly, they express the chances of landing on the black parts of various spinners by marking chance ladders.

Goals

Students will:

- describe chance in everyday language;

- estimate chances in percents from 0% to 100%;

- find chances, in percents, fractions, or ratios, for simple situations.

Pacing

- approximately three or four 45-minute class sessions

About the Mathematics

This section introduces the "chance" ladder model to build a basic understanding that the chance of an event's occurring ranges from 0% to 100%. To estimate chance, students first determine on which rung of the ladder each event belongs and then estimate the percent that corresponds to that rung on the ladder. There are several informal ways of describing chances, such as using percents on chance ladders and using ratio terminology (a one-out-of-six chance). These are precursors to using exact percents and fractions. Percents, ratios, and fractions were introduced in the preceding grade 5/6 units *Per Sense* and *Some of the Parts*. At the end of this section, the informal and exact ways of describing chance are summarized.

Materials

- Student Activity Sheets 1–3, pages 101–103 of the Teacher Guide (one of each per student)
- drawing paper, pages 31 and 33 of the Teacher Guide (four sheets per student)
- black crayons, pages 37 and 41 of the Teacher Guide (one per student)

Planning Instruction

This section is a continuation of the previous section, so no introduction is necessary.

Students may work individually on problems 1 and 11 and in pairs or individually on problems 12, 13, 18, and 19. They may work on the remaining problems in pairs or in small groups.

There are no optional problems in this section.

Homework

Problems 1 (page 28 of the Teacher Guide), 4 and 5 (page 32 of the Teacher Guide), 10 (page 36 of the Teacher Guide), 17 (page 44 of the Teacher Guide), and 18 and 19 (page 46 of the Teacher Guide) can be assigned as homework. After students complete Section B, you may assign appropriate activities from the Try This! section, located on pages 39–42 of the *Take a Chance* Student Book. The Try This! activities reinforce the key math concepts introduced in this section.

Planning Assessment

- Problem 4 may be used to informally assess students' ability to describe chance in everyday language.
- Problem 5 may be used to informally assess students' ability to estimate chances in percents from 0% to 100%.
- Problems 11, 13, and 15 may be used to informally assess students' ability to find chances, in percents, fractions, or ratios, for simple situations.

B. WHAT'S THE CHANCE?

Up and Down Events

Sometimes it is difficult to predict whether or not an event will take place. Other times you know for sure.

1. Use **Student Activity Sheet 1.** Put a check in the column that best describes your confidence that each event will take place.

	Statement	Sure It Won't	Not Sure	Sure It Will
A.	You will have a test in math sometime this year.			
B.	It will rain in your town sometime in the next four days.			
C.	The number of students in your class who can roll their tongues will equal the number of students who cannot.			
D.	You will roll a "7" with a normal number cube.			
E.	In a room of 367 people, two people will have the same birthday.			
F.	New Year's Day will come on the third Monday in January.			
G.	When you toss a coin once, heads will come up.			
H.	If you enter "2 + 2 =" on your calculator, the result will be 4.			

Solutions and Samples
of student work

1. Answers will vary for some statements. Sample student responses:

   a. Sure it will.

   b. Not sure. (This may depend on the time of year or where students live.)

   c. Not sure.

   d. Sure it won't.

   e. Sure it will. (For 366 people, the answer would be "not sure" because each person could have his or her birthday on a different day, including February 29 during leap years. For 367 people, two birthdays must fall on the same day.)

   f. Sure it won't.

   g. Not sure.

   h. Sure it will.

Hints and Comments

Materials Student Activity Sheet 1 (one per student)

Overview Students estimate whether or not eight given events will occur.

About the Mathematics Students gain a basic understanding of the concept of *chance* by estimating chances before describing exact chances using percents, fractions, or ratios.

Students base their chance estimates on the general idea that some events are sure to happen, some are sure not to happen, and all the other possibilities are between these two extremes. On the next page, students think about various intermediary positions.

Planning You may assign this first problem as homework in advance, and begin this section with a class discussion of students' answers.

Comments about the Problems

1. Homework This problem can be assigned as homework. Ask students to explain their answers. Some answers may reflect unusual situations (for example, answers to problems **a, b,** and **c**). Allow students to explain.

   e. Remind students that a year has 365 days and a leap year has 366 days. So, there are 366 different days for a birthday.

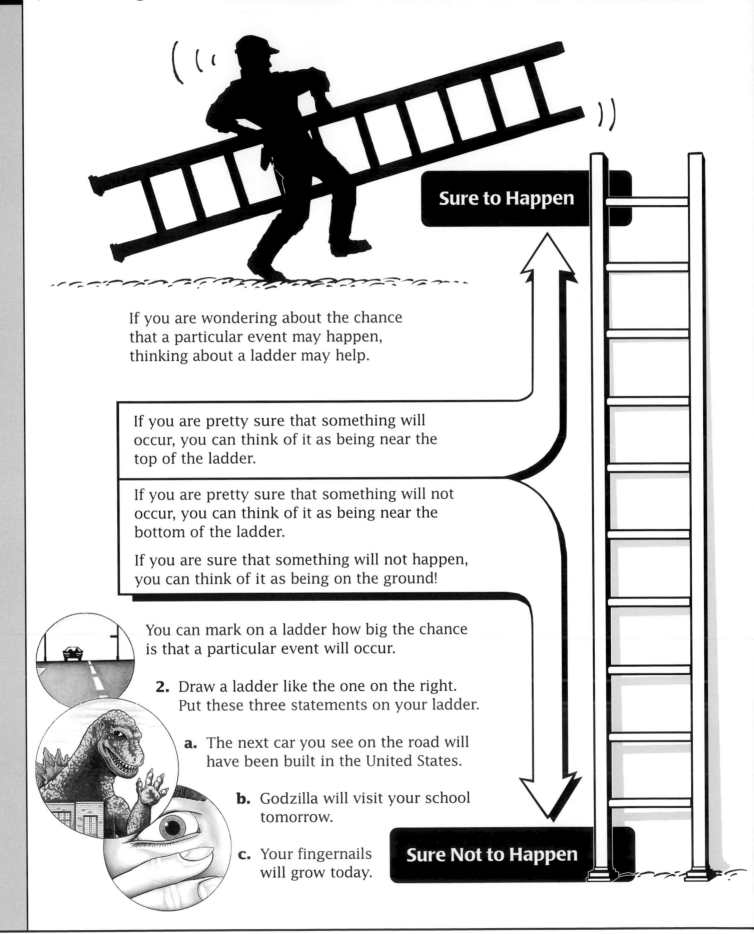

Sure to Happen

If you are wondering about the chance that a particular event may happen, thinking about a ladder may help.

If you are pretty sure that something will occur, you can think of it as being near the top of the ladder.

If you are pretty sure that something will not occur, you can think of it as being near the bottom of the ladder.

If you are sure that something will not happen, you can think of it as being on the ground!

You can mark on a ladder how big the chance is that a particular event will occur.

2. Draw a ladder like the one on the right. Put these three statements on your ladder.

a. The next car you see on the road will have been built in the United States.

b. Godzilla will visit your school tomorrow.

c. Your fingernails will grow today.

Sure Not to Happen

2. Ladders will vary. Sample ladder:

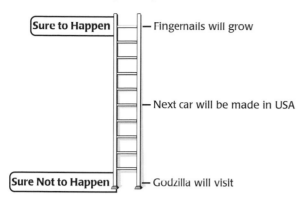

Materials drawing paper (one sheet per group)

Overview Students estimate the chances of three events occurring by placing each event in the appropriate place on a chance ladder.

About the Mathematics The addition of 10 rungs on the ladder makes it necessary for students to refine the category "Not Sure" of problem **1.** They are still allowed to express chances in terms of being "pretty sure" that something will or will not happen in order to estimate the placement of each event on the chance ladder.

Planning Students may work on problem **2** in pairs or in small groups.

Comments about the Problems

2. The 10 rungs of the ladder correspond to the 10% benchmark and its multiples, with 0% at ground level and 100% at the highest rung. Do not point this out to students yet. This concept is introduced on the next page.

3. Now go back to the table on page 11 and put the statements from the table on one ladder. Explain why you put the statements where you did.

Sure to Happen

4. Put the following statements about chance on a ladder:
"I'm sure it will happen."   "There's a 50-50 chance."
"That's unlikely."   "It's very likely to happen."
"It probably will."   "There's no way it will occur."
"There's a 100% chance."   "It seems very unlikely."
"There is a 0% chance."   "It could happen."

Sure Not to Happen

Match 'Em Up

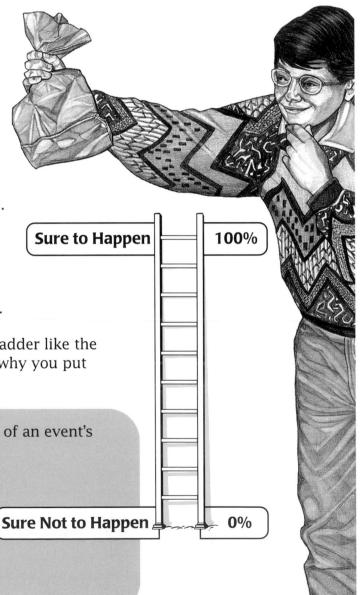

Dan is doing an experiment. He has a bag holding pieces of paper of equal size, numbered 1 to 20. He is going to pick a number from the bag. Here are some possible outcomes for the number he will pick:

    a. It will be even.

    b. It will be divisible by five.

    c. It will be a 1 or a 2.

    d. The digits in the number will add up to 12.

    e. It will be smaller than 16.

Sure to Happen   **100%**

5. Put the five statements on a ladder like the one on the right and explain why you put them where you did.

These ladders show that the chance of an event's happening is between 0% and 100%.

- Events you are sure are going to happen will be at the top.

- Events you are not sure about will be somewhere in between.

- Events you are sure will not happen will be at the bottom.

Sure Not to Happen   **0%**

3. Ladders will vary. Sample student response:

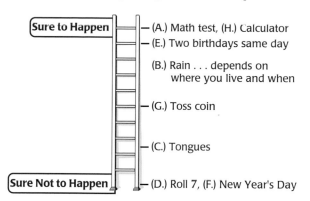

Sure to Happen
— (A.) Math test, (H.) Calculator
— (E.) Two birthdays same day
  (B.) Rain . . . depends on
       where you live and when
— (G.) Toss coin
— (C.) Tongues

Sure Not to Happen
— (D.) Roll 7, (F.) New Year's Day

4. Ladders will vary. Sample student response:

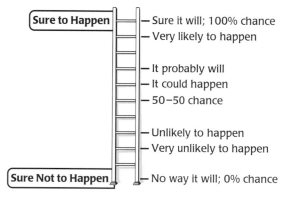

Sure to Happen
— Sure it will; 100% chance
— Very likely to happen
— It probably will
— It could happen
— 50–50 chance
— Unlikely to happen
— Very unlikely to happen

Sure Not to Happen
— No way it will; 0% chance

5. Ladders will vary. Sample student response:

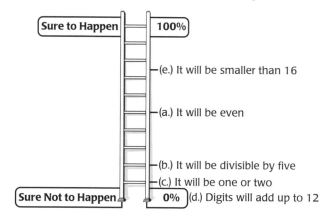

Sure to Happen
100%
—(e.) It will be smaller than 16
—(a.) It will be even
—(b.) It will be divisible by five
—(c.) It will be one or two

Sure Not to Happen
0% (d.) Digits will add up to 12

Materials drawing paper (three sheets per group)

Overview Students estimate the chances of events occurring by placing each event in an appropriate position on a chance ladder. The ladder shows that the chance that an event will occur is between 0% and 100%.

About the Mathematics The use of chance ladders builds a basic understanding that chances range between 0% and 100%. Students first learn to describe chance informally by placing statements on a chance ladder at a height that represents the chance in percent. Another informal way to describe chance is by using ratio terminology (for example, one out of six). Students will use this method later in the section.

Planning Students may continue to work on problems **3–5** in pairs or in small groups. You may use problems **4** and **5** as assessments or assign them as homework. Discuss students' explanations for their answers to problems **3** and **5.**

Comments about the Problems

4. Informal Assessment This problem assesses students' ability to describe chance in everyday language. It may also be assigned as homework.

   Students should be able to use the chance ladder and understand that chances occur in a range. You may point out to students that the ladder has 10 rungs, and ask them whether or not this is useful. If you notice that students are having difficulty with the verbal statements, ask them to draw a ladder and use their own words to describe the chances that correspond to the different heights.

5. Informal Assessment This problem assesses students' ability to estimate chances in percents from 0% to 100%. It may also be assigned as homework. The purpose of this problem is only to order the chances. Allow students to make global comparisons. For example, have them compare **b** and **c.** [There are more numbers that correspond to statement **b** than numbers that correspond to statement **c,** so the chance that a number will be a 1 or a 2 is lower.] Students have the option of calculating exact percents here, but do not make it mandatory. (You may wish to have students figure out the percents later. The actual percents are 50%, 20%, 10%, 0%, and 75%.)

HILLARY IS WALKING TO THE SCIENCE LAB CARRYING HER PET BULLFROG, NEWTON.

FROG SKELETON

NEWTON, IN FEAR FOR HIS LIFE...

...JUMPS OUT OF HIS AQUARIUM AND HOPS OFF AS FAST AS HIS LITTLE FEET CAN CARRY HIM.

Hillary finally found Newton. He was sitting on this tile:

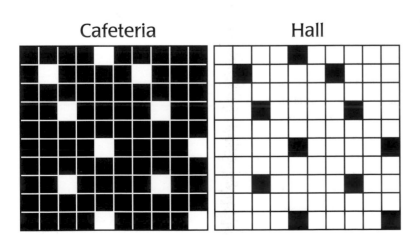

Cafeteria   Hall

6. Look at the two floors on the left. Do you think Hillary found Newton in the cafeteria or the hall? Explain.

7. What if, instead, Newton was sitting on this tile:

Is it likely that he was on the same floor?

Solutions and Samples
of student work

6. Hillary probably found him in the cafeteria. There are more black tiles on the cafeteria floor, so there is a greater chance that the tile is from this floor.

7. No. Hillary probably found him in the hall, since there are more white tiles in that room.

Hints and Comments

Overview Using the context of a frog that jumps on a tile floor, students describe the chances that the frog will land on a black tile by comparing the numbers of black and white tiles.

About the Mathematics The floors with black and white tiles give students visual support for estimating chances. This context prepares students to use informal ratio terminology, such as "one out of four."

Planning Students may continue to work on problems **6** and **7** in pairs or in small groups. Discuss their answers to these problems as a class.

Comments about the Problems

6–7. Some students may respond that the frog could have been found in either room. Ask students: *In which room is it more likely that the frog might be on a black tile or white tile?* [Newton will more likely be on a black tile in the cafeteria and will more likely be on a white tile in the hall.]

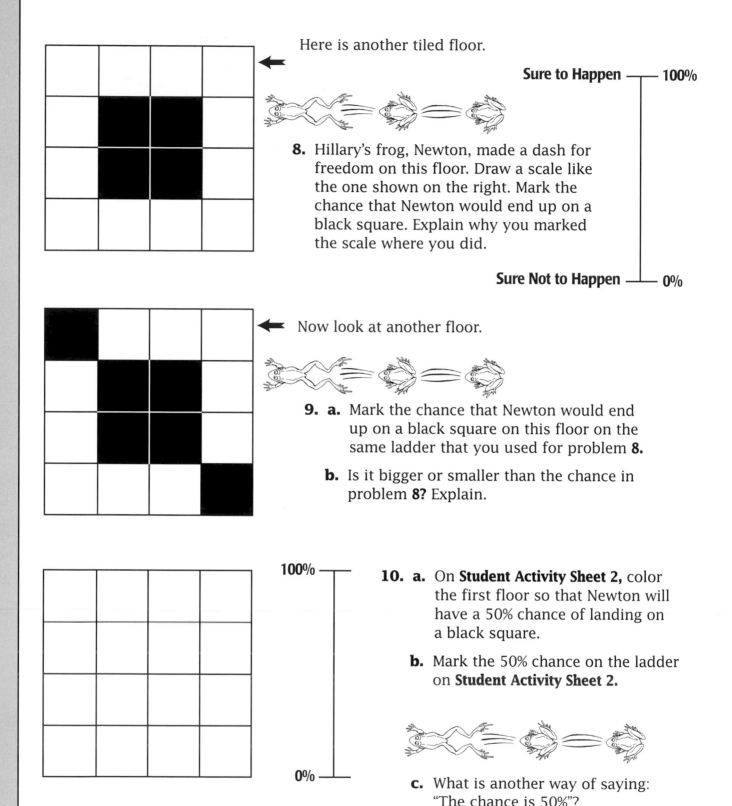

Here is another tiled floor.

Sure to Happen ——— 100%

8. Hillary's frog, Newton, made a dash for freedom on this floor. Draw a scale like the one shown on the right. Mark the chance that Newton would end up on a black square. Explain why you marked the scale where you did.

Sure Not to Happen ——— 0%

Now look at another floor.

9. **a.** Mark the chance that Newton would end up on a black square on this floor on the same ladder that you used for problem **8.**

   b. Is it bigger or smaller than the chance in problem **8?** Explain.

100% ——

10. **a.** On **Student Activity Sheet 2,** color the first floor so that Newton will have a 50% chance of landing on a black square.

   b. Mark the 50% chance on the ladder on **Student Activity Sheet 2.**

0% ——

   c. What is another way of saying: "The chance is 50%"?

8. Answers may vary. Sample drawing:

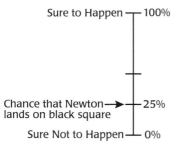

9. a. Answers may vary. Sample drawing:

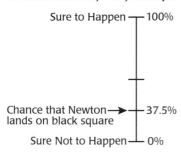

b. Bigger. In the floor for problem **8,** only four out of the sixteen squares are black. In problem **9,** six out of the sixteen squares are black. Since there are more black squares in problem **9,** it is more likely Newton will land on a black square.

10. a. Floors will vary. Accept any solution that shows one-half of the squares black and one-half of the squares white. You may want students to compare their solutions.

Sample solutions:

b.

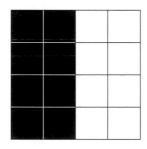

c. Answers will vary. Sample responses:
- one out of two chances, or
- a 50-50 chance.

Materials Student Activity Sheet 2 (one per student), black crayons (one per student)

Overview Students use percents to estimate the chances that a frog will land on a black tile on different black-and-white tile floors.

About the Mathematics The chance ladder is now reduced to a vertical scale line without markings for multiples of 10 percent. The main reference points on the scale are 0%, 50%, and 100%.

Planning Students may continue to work in pairs or in small groups on problems **8–10.** You may assign problem **10** for homework. Remind students to save their copy of Student Activity Sheet **2** to use again in problem **12** on page 17 of the Student Book.

Comments about the Problems

8. Discuss students' explanations to this problem. Some students may reason that the frog has a 50% chance of landing on a black tile since the floor is made up of only black and white tiles. If no students offer this explanation, ask them to react to this statement during the discussion.

8–9. Students do not have to compute the exact percents. They should, however, be able to estimate the chance by placing each event in an appropriate position on the ladder. Informal answers are still acceptable here. If students say that the frog will not land on a black square because it is too far away, remind students that the frog has the ability to jump on any square.

10. Homework This problem can be assigned for homework.

11. a. For the floor in problem **8,** you can say that the chance of landing on a black square is 4 out of 16. Explain this.

b. Jim says,

> THAT'S THE SAME AS 1 OUT OF 4.

Do you agree? Explain.

c. Here is the floor from problem **9.** What is the chance of landing on a black square on this floor?

11. a. Explanations may vary. Sample explanation:

There are 4 black squares and 16 total squares, so the chance of landing on a black square is 4 out of 16.

b. Yes. The ratio 1 out of 4 is the same as 4 out of 16. They are equivalent ratios.

c. 6 out of 16, or 3 out of 8

Overview Students use informal ratio terminology to express the chances of the frog landing on a black square on different black-and-white tile floors.

About the Mathematics On the previous page, percent estimates were made by comparing the numbers of black and white squares. On this page, ratio terminology, such as "one out of six," is introduced. To describe chance using this terminology, it is necessary that the black tiles be compared with the total number of squares (the total possibilities). The context gives students the opportunity to make connections with the parking lot context in the grade 5/6 unit *Per Sense*, in which students describe the fraction of occupied spaces using the same terminology.

Planning You may ask students to solve problem **11** individually. This will give you the opportunity to see how well each student understands and uses ratio terminology. You may decide to use this problem or problem **13** on the next page for assessment.

Comments about the Problems

11. Informal Assessment This problem assesses students' ability to find chances, in percents, fractions, or ratios, for simple situations.

For part **b,** you can show how the picture validates $\frac{1}{4}$ by partitioning and sliding the squares. This can be done in two different ways:

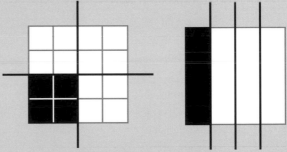

In part **c,** students are asked to find the exact chance of landing on certain tiles for the first time. All of the previous problems could be answered informally. Some students may reduce the ratio 6 out of 16 to 3 out of 8. However, this is not necessary.

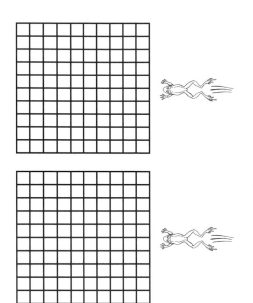

12. a. Color the second floor on **Student Activity Sheet 2** so that Newton's chance of landing on a black square is 1 out of 5.

b. Now color the third floor on **Student Activity Sheet 2** with any pattern of black and white tiles. What is the chance that Newton will land on a black tile on the floor you made?

13. If you had a black-and-white tile floor, explain how you would find the chance that a frog would land on a black square.

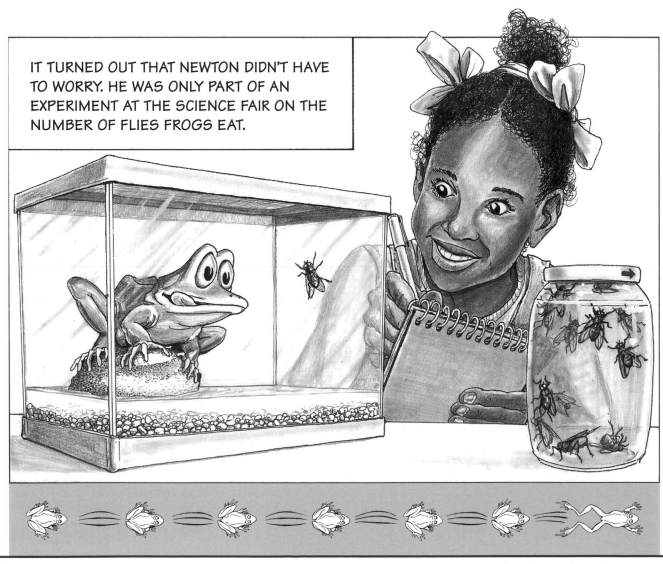

IT TURNED OUT THAT NEWTON DIDN'T HAVE TO WORRY. HE WAS ONLY PART OF AN EXPERIMENT AT THE SCIENCE FAIR ON THE NUMBER OF FLIES FROGS EAT.

12. a. Answers will vary. Accept any solution in which 20% or $\frac{1}{5}$ of the squares are black.

Sample solutions:

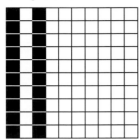 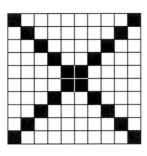

b. Answers will vary depending on individual designs. The chance will be the number of black squares out of the total number of squares. Students might also express this as a fraction.

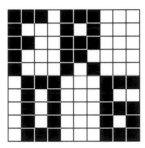 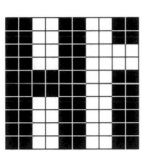

44 black squares

percent: 44%

fraction: $\frac{44}{100}$ or $\frac{11}{25}$

ratio: 44 out of 100 or 11 out of 25

60 black squares

percent: 60%

fraction: $\frac{60}{100}$, $\frac{6}{10}$, or $\frac{3}{5}$

ratio: 60 out of 100, 6 out of 10, or 3 out of 5

13. You could count the number of black squares and the total number of squares in the floor. The chance of a frog's landing on a black square is the ratio of the black squares to the total number of squares.

Materials Student Activity Sheet 2 (one per student), black crayons (one per student)

Overview In the previous problems, students described chances based on the numbers of black and white tiles on given floors. Now they will work backwards by using a given chance to create a tile pattern for a floor. Students also generalize how to find the chance for landing on a black square.

About the Mathematics Students may use what they learned about ratios and proportions in the units *Some of the Parts* and *Per Sense* to solve these problems. Several strategies can be used to solve problem **12:**

• consecutively coloring one out of every five tiles,

• coloring one row out of every five rows (or two out of every ten rows),

• computing how many tiles need to be colored before coloring the tiles.

Equivalent ratios are generated with all of these strategies. Some of the equivalent ratios are shown in the ratio table below.

Number of Black Tiles (or Rows)	1	2	20
Total Number of Tiles (or Rows)	5	10	100

This problem may also reinforce students' understanding about the relationship between ratios, fractions, and percents. For example, the ratio *one out of five* is equivalent to the fraction $\frac{1}{5}$ and to 20%.

Planning Be sure students still have their copies of Student Activity Sheet 2. They may work on problems **12** and **13** individually. You may want to use problem **13** for assessment. Discuss students' strategies and solutions to problem **12** with the whole class.

Comments about the Problems

12. a. Ask students to share their strategies and computations before they begin coloring.

b. Ask students to show how they found the chance of landing on a black tile.

13. Informal Assessment This problem assesses students' ability to find chances, in percents, fractions, or ratios, for simple situations.

Spinners

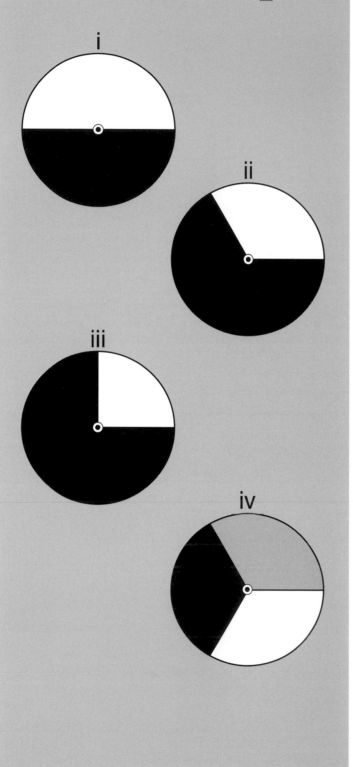

14. Study the spinners on the left.

    a. Can you use spinner i to make fair decisions? Explain your answer.

    b. Can you use spinner ii, iii, or iv to make a fair decision?

    c. Draw a new spinner—different from i, ii, iii, and iv—that can be used to make fair decisions.

15. a. Draw a chance ladder in your notebook. For each spinner on the left, mark the ladder to show the chance of landing in the black part.

    b. Use a method other than a ladder to express the chance of landing in the black part of each spinner.

14. a. Yes, if two people or choices are involved. Since there is an equal chance of landing on either section, the spinner is fair.

b. You cannot use spinners ii and iii to make a fair decision between two people. Spinner iv may be used to make a fair decision with three people.

c. Spinners will vary. Student spinners will probably have sections with equal areas. Sample solutions:

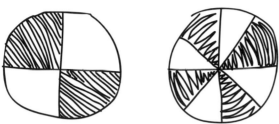

15. a.

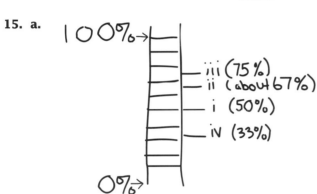

b. Answers will vary. Students may express the chances using percents, ratios, or fractions.

Sample solutions:

i: 50%, one out of two, or $\frac{1}{2}$

ii: about 67%, two out of three, or $\frac{2}{3}$

iii: 75%, three out of four, or $\frac{3}{4}$

iv: about 33%, one out of three, or $\frac{1}{3}$

Overview Students decide whether or not different spinners can be used to make fair decisions. They represent the chance for landing on the black part of each spinner with a chance ladder and express the chance as a percent, a fraction, or a ratio.

About the Mathematics Technically, whether or not spinners can be used to make fair decisions depends upon what is being decided. For example, if you want to give outcomes different chances, a spinner with unequal central angles might be considered fair. It is not important to discuss this with students at this time.

Planning Students may work on problems **14** and **15** in pairs or in small groups. You can use problem **15** for assessment.

Comments about the Problems

14. In this problem, students use spinners to make fair decisions, as was done in Section A. If some students are having difficulty, refer them to that section.

15. Informal Assessment This problem assesses students' ability to find chances, in percents, fractions, or ratios, for simple situations.

Students should be able to justify their answers using fractions or percents. Encourage them to express their pictorial answer in part **a** as a ratio, fraction, or percent if they have not already done so.

Jim made this spinner and colored in this floor. Jim says,

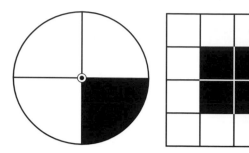

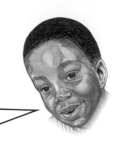

THE SPINNER AND THE FLOOR GIVE THE SAME CHANCE FOR LANDING IN THE BLACK PART.

16. Do you agree with Jim? Explain.

Activity

17. Look in the newspaper for statements about chance. Put these statements on a chance ladder. Bring the ladder to school and explain why you decided to place the statements where you did. Here are some examples to help you.

Baseball Update

Chances for a Run at Division Title Slim

By Mel Bergman
of The Reporter staff

It was no surprise that the Cal-away CooCoos' manager Regg Loopendorf refused to give a statement regarding his team's

even though the seri led to a new definitio Future chances for a can only hope that it

Cease-Fire May End Soon

and that world will end all to bring... and that will return the focus that...

UPL
Reports that a new attempts to bri

Home Buyer's Guide

THIS MAY BE YOUR LAST CHANCE TO BUY A NEW HOME ON SILVER LAKE

Site rating guide
★★★★

LATEST LUXURIOUS LISTINGS

The last twenty homes will go on sale this weekend.

The increase in the number of families wanting to settle in this spectacular area has increased dramatically over the last few weeks. The few

the range of views and the availability of easy access to many of the recreational outlets bring a new meaning to the term "Land of Dreams"

16. Jim is right. There is a one-out-of-four chance of landing on black using either the spinner or the tile floor.

17. Ladders will vary. Students might find chance statements from advertisements, sports articles, poll results, and so on.

Sample chance ladder:

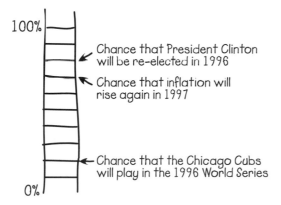

100% — Chance that President Clinton will be re-elected in 1996

Chance that inflation will rise again in 1997

Chance that the Chicago Cubs will play in the 1996 World Series

0%

I think that President Clinton has about an 80 percent chance of being re-elected in 1996 because according to the latest poll in the newspaper article I read, 75 percent of the people polled said they were going to vote for him in the November election. So I positioned this statement at 80 percent on the chance ladder.

According to the newspaper article that I read, inflation will continue to slowly rise over the next five years. So, I positioned this statement at 70 percent on the chance ladder.

According the newspaper article that I read, the Chicago Cubs have not played in the World Series for many years. So I positioned this statement at 20 percent on the chance ladder.

Overview Students compare the chance of landing on the black part of a spinner to the chance of landing on the black tiles on a floor. Students then share with the class chance ladders they constructed to represent newspaper headlines about chance.

Planning Students may do problems **16** and **17** in pairs or small groups. You may assign problem **17** for homework or as a project. If students do not have newspapers at home, you may want to make newspapers available to students in class.

Comments about the Problems

16. This problem makes explicit connections between chances on a spinner and chances on a tile floor. You may ask students to design another spinner and floor that have equal chances.

17. Homework This problem may be assigned as homework. The point of the activity is to get a sense of the language of chance in everyday occurrences.

Summary

In this section, you saw different ways of expressing chances. You have seen that the chances on a ladder can be expressed with percents. If you are sure that something will happen, you can say the chance is 100%. If you are sure that it will not happen, the chance is 0%. Chances can also be expressed with fractions. You can make a chance ladder using fractions.

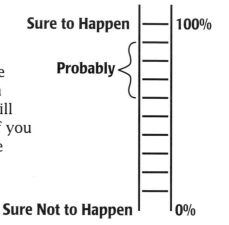

Summary Questions

18. a. What fraction would you use to represent a 50-50 chance?

b. Put some other fractions where they belong on a chance ladder.

Below are some statements about chances. Some of them belong together; they are just different ways of saying the same thing.

19. On **Student Activity Sheet 3,** connect all statements that say the same thing. One example has already been done.

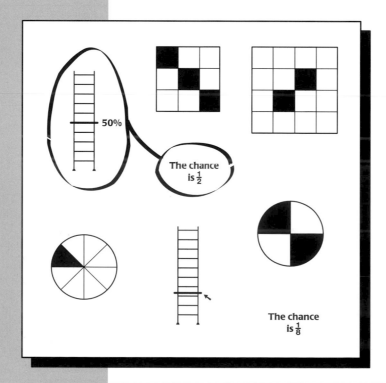

18. **a.** Answers will vary. Accept any fraction that is equivalent to $\frac{1}{2}$.

Sample solutions: $\frac{1}{2}$, $\frac{5}{10}$, or $\frac{50}{100}$

b. Answers will vary. Sample chance ladder:

19.

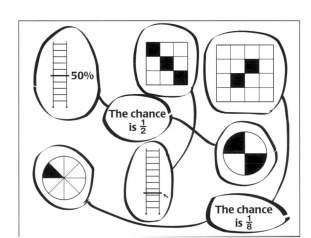

Materials Student Activity Sheet 3 (one per student)

Overview This page summarizes the main concepts of this section. Different ways of describing chance are tied together: chance ladders, percents, and fractions.

Planning After students read the Summary, you may want them to work on problems **18** and **19** individually. You may decide to assign these problems for homework. Discuss students' answers to these problems with the whole class. After students complete Section B, you may assign appropriate activities from the *Try This!* section, located on pages 39–42 of the *Take A Chance* Student Book, for homework.

Comments about the Problems

18. **a. Homework** Some students may still be using the phrase *one out of two.* If so, encourage them to express this as a fraction. Ask them whether or not that ratio is the same as the fraction $\frac{1}{2}$.

b. You may ask your students what fractions belong at *Sure to Happen* and *Sure Not to Happen.*

19. **Homework** This problem may be used as homework. Some students may need the hint that sometimes more than two statements can be connected.

SECTION C. LET THE GOOD TIMES ROLL

Work Students Do

Students investigate what happens when a number cube is rolled many times. They toss coins, tally the outcome, and analyze the results. They observe how the results stabilize as the experiment continues. Students repeat the process in various other situations, such as picking a number from one to four and tossing a cup, a bottle cap, or an eraser.

Goals

Students will:

- estimate chances in percents from 0% to 100%;

- find chances, in percents, fractions, or ratios, for simple situations;

- list the possible outcomes of simple chance and counting situations;*

- understand that in independent trials, the next outcome is not affected by previous outcomes;

- use repeated trials of a single experiment to estimate chances;*

- understand that variability is inherent in any probability situation;

- develop an understanding of the difference between theoretical and experimental probability.

  * *These goals are introduced in this section and will be assessed later in the unit.*

Pacing

- approximately three 45-minute class sessions

Vocabulary

- trial

About the Mathematics

This section addresses the difference between theoretical and experimental chance and the independence of outcomes when tossing a coin or rolling a number cube. Theoretical chance is based on considerations about the possible outcomes of a chance experiment and can be determined *before* the experiment is conducted. For example, the theoretical chance of landing on heads or tails when tossing a coin is 50% for each outcome. The theoretical chance for rolling one of the six numbers on a number cube is $\frac{1}{6}$. Experimental chance deals with trial situations in which the chances of each outcome cannot be predicted. For example, it is difficult to predict the chances for landing on a particular side when irregularly shaped objects are rolled.

Although one can predict theoretical chance, the outcome of a single trial is unpredictable. Only after a large number of repeated trials will the outcomes stabilize and fall into a predictable pattern.

For experimental chance situations, the distribution of the outcomes of a large number of repeated trials can be used to estimate the probability of each outcome.

Materials

- number cubes, page 51 of the Teacher Guide (one per group of students)
- coins, page 55 of the Teacher Guide (one per group of students)
- students' work from problem 19 in Section A, page 59 of the Teacher Guide (one per student)

Planning Instruction

You may start this section by asking your class how many of them are familiar with number cubes. If a substantial number of students have never used them before, you may want to have them play a math game that uses number cubes. Make sure, however, that students discuss the game in terms of chance.

If time is a concern, you may want to have the students do one or two of the activities in this section at home. However, discuss the activities before you assign them as homework.

Students may work on most problems in pairs or in small groups.

Problems 12 and 13 are optional. If you decide to use them, work on problem 12 as a whole class, and let students complete problem 13 individually.

Homework

Problems 8–10 (page 54 of the Teacher Guide), the Extension (page 51 of the Teacher Guide), and the Writing Opportunity (page 55 of the Teacher Guide) can be assigned for homework. After students complete Section C, you may assign appropriate activities from the Try This! section, located on pages 39–42 of the *Take a Chance* Student Book. The Try This! activities reinforce the key math concepts introduced in this section.

Planning Assessment

- Problem 7 can be used to informally assess students' understanding that variability is inherent in any probability situation and their understanding of the difference between theoretical and experimental probability.

- Problem 14 can be used to informally assess students' ability to find chances, in percents, fractions, or ratios, for simple situations.

- Problem 15 can be used to informally assess students' understanding of the fact that in independent trials, the next outcome is not affected by previous outcomes and their understanding of the difference between theoretical and experimental probability. It also assesses their ability to estimate chances in percents from 0% to 100%.

C. LET THE GOOD TIMES ROLL

Chancy Business

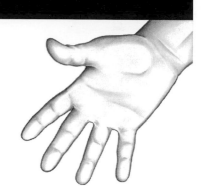

If you roll a number cube one time, the chance that you will roll a 6 is the same as the chance of rolling a 5, a 4, a 3, a 2, or a 1.

1. If you roll a number cube 30 times, about how many times would you expect to roll a 6?

Activity

2. Make a table like the one shown on the right. Roll a number cube 30 times. Tally the number that comes up for each roll.

Number Rolled	Number of Times It Came Up
1	
2	
3	
4	
5	
6	

3. Did what happened differ from what you expected would happen? How?

4. What do you think would happen if you increased the number of rolls to 60?

1. You would expect to roll a 6 about five times.

2. Answers will vary. While on average each number will come up five times, this will almost never happen in just 30 rolls. There will be too much variability.

Sample student table:

Number Rolled	Number of Times It Came Up
1	||\\\|
2	₶
3	₶ |
4	||||
5	₶ ||||
6	||

3. Answers will vary. With only 30 rolls, some students may have expected that the results would be unbalanced, while other students may have expected that each number would come up five times.

4. Answers will vary, but should suggest that as the number of rolls increases, the results should become more balanced. For example, in 60 rolls, one would expect each number to come up about 10 times.

Materials number cubes (one per group of students)

Overview Students predict the outcome of a chance experiment. They roll a number cube 30 times and tally the results. Then they compare their results to their predictions.

About The Mathematics Students explore the differences between theoretical chance and experimental chance. Based on theoretical arguments, one can predict the outcomes when rolling a number cube or tossing a coin. But when actually conducting experiments, there may be a lot of variability in the outcomes. The actual outcomes may differ somewhat from theoretical chances.

The problems on this page introduce two important mathematical concepts:
• in independent trials, the next outcome is not affected by previous outcomes, and
• by conducting repeated trials of an experiment, the outcomes will approach the expected outcomes.

Planning Remind students that if a number cube is fair, each number has an equal chance of being rolled. Have students complete problems **1–4** in pairs or in small groups.

Comments about the Problems

1. Let students make their own predictions. If they are having difficulty, ask them how many times they would expect to roll each of the outcomes 1–6.

2. You may need to demonstrate how to record data on a tally sheet if students have had no previous experience using one.

3. You might ask students to explain why the results of the experiment differed from the expected outcome as determined in problem **1.** Some students may reason that they would need to conduct more than 30 trials to come closer to the expected outcome.

4. Students can investigate what happens in 60 trials by combining the results of two tables.

Extension You may want to combine the tables of the whole class and see whether or not the results stabilize. Discuss the results with the class. With more data, the actual outcome of each number should come closer to expected theoretical outcomes than the results of any one table.

Nina rolled a number cube. She recorded the results in the table on the right.

Number Rolled	Number of Times It Came Up
1	///
2	THL THL /
3	THL THL
4	THL THL /
5	THL ///
6	THL THL ///

5. a. How many times did Nina roll the number cube?

b. Nina says,

THE CHANCE OF ROLLING A 1 ON THE NEXT ROLL IS GREATER THAN THE CHANCE OF ROLLING A 6.

Do you think that she is right?

6. a. Robert rolled a number cube six times. Do you think that he rolled a 4? Explain.

b. Then Robert rolled the number cube 20 times more. Do you think that he rolled a 4 this time?

Now We're Rolling!

Hillary rolled a number cube many times as part of an experiment.

Number Rolled	1	2	3	4	5	
Number of Times It Came Up	44	36	37	41	39	

7. Unfortunately, Hillary's pen leaked and covered up the number of times that 6 came up. What do you think is written under the spill? Explain.

5. a. Nina rolled the number cube 56 times.

   b. Nina is wrong. Even though she did not get as many ones as she had expected, that does not affect the chance of rolling a 1 on the next roll.

6. a. Answers will vary. Possible response:

   You can't be sure that Robert rolled a 4, since he rolled the number cube only six times.

   b. Answers will vary. In this case, it is almost certain he has thrown a 4 but, again, not absolutely certain.

7. Answers will vary. However, since the rolls for the other numbers ranged from 36 to 44, it is reasonable to expect the number of sixes also to be in the high 30s or low 40s. The number of sixes could be outside this range, however.

Overview Students solve more chance problems involving number cubes and reason about the possible outcomes.

About the Mathematics Students compare theoretical chances to experimental chances in problems using number cubes. Problems **5** and **6** focus on the fact that the outcome of each roll is independent of the outcomes of previous rolls. Problem **7** deals with variability in the results of a chance experiment.

Planning Students may continue to work in pairs or small groups on these problems. Discuss problem **5b** with the whole class. Students should realize that the outcome of each roll is not affected by outcomes on previous rolls. You may want to have them continue to experiment with rolling a number cube to see that this is actually so.

Comments about the Problems

5. a. Make sure students realize that counting tallies is easier if you count first by fives or by tens, and then count the single tallies.

   b. This may be a challenging problem for some students. Looking at the table, they may erroneously reason that since so few ones have been rolled so far, there is a greater chance of rolling a one. It may help to state that on each roll, the number cube does not remember what happened before, so the chances for each outcome remain unchanged.

6. a. You may have students replicate Robert's experiment a few times. It is likely that the number 4 may not occur even in six rolls.

   b. Again, you may have students actually roll a number cube 20 more times to support their reasoning.

7. Informal Assessment This problem assesses students' understanding that variability is inherent in any probability situation and their understanding of the difference between theoretical and experimental probability.

This problem focuses on the fact that outcomes tend to approach theoretical probabilities after repeated trials. In this case, each outcome will occur about the same number of times. Try to have students *estimate* the number of sixes based on this kind of consideration. Making calculations is not necessary since no *exact* answer is possible.

Tossing and Turning

During World War II, English mathematician John Kesrich was locked in a cell. He had a coin with him and decided to do an experiment to pass the time. While in the cell, he tossed the coin 10,000 times and recorded the results.

Here is the start of a chart he might have made.

Number of Tosses	Total Number of Heads
1	0
2	0
3	1
4	1
5	2
6	2
7	3
8	3
9	4

8. a. Was the first toss a head?

   b. On which toss did the mathematician get heads for the first time?

   c. How many tosses did it take to get three heads?

   d. How many tails were thrown after eight tosses?

9. About how many heads do you think would come up after 10,000 tosses?

10. How does the percent of heads change as the number of coin tosses increases?

Activity

11. a. If you toss a coin 30 times, how many times do you expect heads to come up?

Toss a coin 30 times. Tally the results in a chart like the one below.

H	T

   b. Did your results match what you predicted?

   c. Combine your results with those of everyone in the class. How do the class results compare with your individual results?

8. a. no

   b. on the third toss

   c. seven tosses

   d. five tails

9. about 5,000 heads

10. The percent of heads should approach 50% as the number of coin tosses increases.

11. a. Heads should come up about 15 times.

   b. Answers will vary. It is quite unlikely that every student will get exactly 15 heads and 15 tails. However, students' actual results should be close to these expected results.

   c. Students should notice that the class results are more in line with the expected results of a 50-50 chance of heads or tails coming up.

Materials coins (one per group of students)

Overview Students investigate the differences between theoretical and experimental chances within the context of tossing a coin.

About the Mathematics Although the mathematical concepts dealt with on this page are essentially the same as with the number cube, the context of tossing a coin is different. In a coin toss, the two outcomes are equally likely to occur. The chances are expressed using percents. Since the chances of tossing a head or a tail are each $\frac{1}{2}$ or 50%, students should have few problems working with percent.

The table in problem **8** differs from other tables used in this section. It is a cumulative table that shows the total number of heads over time. On each roll, the total number of heads either stays the same or increases by one. By comparing the total number of heads to the total number of rolls, the percent of heads can be estimated or calculated. The percent of heads occurring approaches 50% as the number of rolls increases.

Planning Students may continue to work in pairs or in small groups on problems **8–11.** You may want to discuss the table in problem **8** with the whole class to ensure that students are able to read it. After they finish problem **11,** you might have a short class discussion about the fact that experimental outcomes tend to approach theoretical ones with repeated trials.

Comments about the Problems

   8–10. Homework These problems may be assigned as homework.

   8. Before students begin this problem, have them study the table and explain the changes from one row in the table to the next.

   10. Before students answer this problem, ask them how to estimate the percent of heads for each row in the table.

   11. Make sure students understand how to fill in the table.

Writing Opportunity For problem **11c,** you may have students write a short paragraph in their journals explaining what happens to the percent of heads when the number of tosses increases from 30 to 60.

As you toss a coin many times, the percent of heads approaches 50%, or $\frac{1}{2}$.

On any single toss, though, you cannot tell whether a head or a tail will come up.

Although you cannot predict a single event, if you repeat an experiment many times, a pattern can appear.

Think B4 You Act

1 2 3 4

Pick a number from 1 to 4 and write it down on a piece of paper.

12. a. If every student in the class writes down one number, how many times do you expect each number to be picked?

   b. Count all the 1s, 2s, 3s, and 4s selected and put the information in a table. Look at the results. Is this what you expected? Why or why not?

13. If you were a game show host and wanted to put a prize behind one of four doors, where would you put the prize? Give a reason for your choice.

12. a. Answers will vary. You would expect all the numbers to have an equal chance, but three tends to be picked most often, followed by two, then one or four.

b. Answers will vary. The results are probably not what students expected because "random" number picks are not really random, since people tend not to pick the numbers on either end of a given range.

13. A game show host should put the prize where the chance of its being picked is lowest, either behind door one or door four.

Overview The three main concepts dealt with in this section are summarized here. Students discover that outcomes of chance experiments do not always approach theoretical possibilities.

About the Mathematics Students experiment with picking a number from one to four; the chances seem to be predictable, based on theoretical considerations. In fact, however, each number has the same chance of being picked only if the numbers are randomly chosen. Due to the fact that many people have specific preferences for certain numbers and tend to avoid extremes (one and four), people do not choose numbers randomly. So, in this case, the experimental outcomes, even when repeated many times, are not likely to approach theoretical outcomes.

Planning In reviewing the experiment from the previous page, you may want to discuss the text on the top of this page. Discuss what patterns students saw when they repeatedly rolled the number cubes and tossed the coins. These problems are optional. Students may work on Problem **12** as a class and on problem **13** individually.

Comments about the Problems

12. First have each student write down a number from one to four. Tally the results in a table on the chalkboard or overhead projector. The results and the reasons for them can then be discussed. Students' responses may vary. Some students may know from experience that three is a more popular number.

13. Students' explanations are more important than their answers. Some students may reason from the viewpoint of the contestant, rather than the game show host, and put the prize behind the door with their favorite number.

Did You Know? The chance situation described on this page is similar to that of picking a name at random from a phone book by pointing with your finger. The chance that someone would pick the first or last name on the page is not the same as the chance of choosing a name in the middle of the page. You might try this as a class experiment. Perhaps students can come up with more experiments like this.

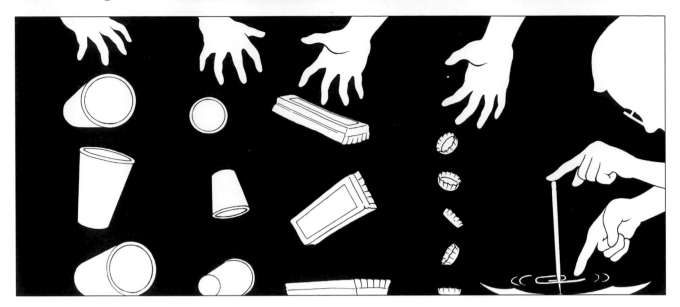

Find the Chance

Back in Section A, problem **19** (page 6), you experimented with these objects:

- a large paper cup

- a small paper cup

- a chalkboard eraser

- a bottle cap

- the spinner on the left

14. Look again at your work on problem **19,** page 6. Based on the 30 throws for your object, estimate the chance that each possibility will occur when you throw the object.

14. Some rough estimates are as follows:

a.

Large Cup Toss	
Top	5%
Side	90%
Bottom	5%

b.

Small Cup Toss	
Top	5%
Side	90%
Bottom	5%

c.

Eraser Toss	
Top	45%
Side	5%
Bottom	45%
Side	5%

d.

Cap Toss	
Top	50%
Bottom	50%

e.

Spinner	
Black	50%
White	50%

Materials Students' work from problem 19 on page 6 of the Student Book.

Overview Students make predictions based on the results of the experiment they conducted with irregularly shaped objects in Section A.

About the Mathematics When rolling number cubes and tossing coins, the theoretical chance of each outcome is known. When tossing or rolling irregularly shaped objects, this is not the case. The only way to get insight into the chances for the different outcomes is through repeated trials with each object. It is not necessary for students to calculate exact chances here. They can order the chances without quantifying, such as *the biggest chance is ...,* or express the chances using numbers, such as *the chances are so many times out of 30.* It is important that students discover that:
- outcomes are not equally likely in all chance experiments,
- you cannot always predict outcomes, and
- an experiment may give insight into the chance distribution.

Planning If students did not complete problem 19 on page 6 of Section A, they will have to do so before beginning this page. Have students work in the same groups with the same objects in order to combine the results of both experiments in one table. Then discuss their findings. Before beginning this problem, discuss the reliability of the chances. Ask students, *Can you be sure that if you throw the paper cup another 30 times, it will land on its side about 27 times?* [no, but probably close to that number] *What do you think will happen to the chances if you throw the paper cup another 30,000 times?* [The chance of the paper cup landing on its top or bottom will be very close to 5 percent and the chance of the paper cup landing on its side will be very close to 95 percent.]

Comments about the Problems

14. Informal Assessment This problem assesses students' ability to find chances, in percents, fractions, or ratios, for simple situations. If students have difficulty, ask them to first order the outcomes from least to most likely. If time permits, allow groups to conduct additional trials to see whether or not their results correspond to the chances they found initially. Discuss students' results, focusing on the fact that not all outcomes are equally likely in experiments with irregularly shaped objects.

Summary

You can find the chance of an event by experimenting with many, many *trials.* In the short run, what happens may not be what you expect. But in the long run, your results will get closer and closer to what you expect.

When tossing a coin or rolling a number cube, each new trial will offer the same chances as the previous one. A coin or number cube cannot remember what side it landed on last.

Summary Questions

15. a. If you toss a coin 10 times and get a head every time, what is the chance of getting a head on the 11th toss?

b. If you roll a number cube over and over again, what do you think will happen to the percent of even numbers that come up as you keep rolling?

16. How many times do you think you have to toss a coin before you can begin to see any patterns in the outcomes?

15. a. 50% or $\frac{1}{2}$

    b. The percent of even numbers will probably get closer to 50%.

16. Answers will vary, but should indicate that the coin needs to be tossed many times. The more times the coin is tossed, the easier it is to see a pattern.

Overview Students read the Summary and make predictions about experiments involving coins and number cubes.

About the Mathematics After discussing the main points of the Summary, you may want to discuss the fact that in experiments with irregularly shaped objects, the chance of each outcome will not be known beforehand.

Planning Students may work on problems **15** and **16** individually or in pairs. You may use problem **15** as an assessment. After students complete Section C, you may assign appropriate activities from the Try This! section, located on pages 39–42 of the *Take A Chance* Student Book, for homework.

Comments about the Problems

15. Informal Assessment This problem assesses students' understanding of the fact that in independent trials, the next outcome is not affected by previous outcomes and their understanding of the difference between theoretical and experimental probability. It also assesses their ability to estimate chances in percents from 0% to 100%.

    b. If students are having difficulty, ask *How many even and odd numbers does a number cube have?* [three even numbers and three odd numbers] *What do you think will happen to the percent of even numbers after repeated trials?* [After repeated trials, the outcomes for even numbers should be about equal to those of odd numbers.] You might also let students roll a number cube many times and tally their results.

16. If students are having difficulty, have them look back at previous problems involving tossing a coin.

SECTION D. LET ME COUNT THE WAYS

Work Students Do

Students explore the role of combinations in probability through different settings, including the possible combinations of boys and girls in families with two children and the possible combinations of different pants and shirts to make one outfit. Students then make predictions by simulating these scenarios with coins and number cubes.

Students use a range of strategies including making pictures, lists, charts, and tree diagrams to represent possible outcomes and to make predictions in an open-hand/closed-fist game, a number cube game, a maze experiment, and a buried treasure activity.

Goals

Students will:

- find chances, in percents, fractions, or ratios, for simple situations;

- list the possible outcomes of simple chance and counting situations;

- understand that in independent trials, the next outcome is not affected by previous outcomes;*

- use repeated trials of a single experiment to estimate chances;

- use tree diagrams to represent simple one-, two-, and three-event situations;

- understand that variability is inherent in any probability situation;

- model real-life situations involving probability.

  * This goal is assessed in other sections of the unit.

Pacing

- approximately four or five 45-minute class sessions

Vocabulary

- tree diagram
- simulation

About the Mathematics

In this section, chance experiments are extended to include finding chances in multi-event situations. For example, a single coin toss is one event. If you are interested in the outcomes of tossing two coins in the same trial (two heads, two tails, or one of each), you need to analyze the two events together. To do this, it is important to count all the possible outcomes for both events. Two effective counting models are introduced in this section: the tree diagram and the counting principle. According to the counting principle, the total number of outcomes in two events is the product of the total number of outcomes in each event. After finding all possible outcomes, the chances of any one outcome are expressed as the ratio of the number of favorable outcomes out of all possible outcomes.

Materials

- Student Activity Sheets 4 and 5, pages 104–105 of the Teacher Guide (one of each per student)
- copies or overhead transparency of the clothes shown on page 28 of the Student Book, page 67 of the Teacher Guide, optional (one copy per student or one transparency)
- pennies, page 65 of the Teacher Guide (two per student or small group of students)
- different-colored number cubes, page 81 of the Teacher Guide (two per pair of students)
- copies of the chessboard, page 84 of the Teacher Guide, optional (one per pair or group of students)

Planning Instruction

You may wish to take a class survey to find the number children and number of boys and girls in each family. You could make a table to see whether or not any conclusions can be drawn from the numbers of boys and girls in each size family. You could relate the class results to those expected from a simulation or from a theoretical analysis.

See the Hints and Comments column on each page for grouping suggestions.

Problems 26 and 29 are optional. If time is a factor, you may omit these problems or assign them as homework.

Homework

Problems 14 (page 74 of the Teacher Guide), 16 (page 76 of the Teacher Guide), 20 (page 78 of the Teacher Guide), and 26–28 (page 82 of the Teacher Guide) can be assigned as homework. The Extensions (pages 69, 71, 79, and 87) and the Writing Opportunity (page 79 of the Teacher Guide) can also be assigned as homework. After students complete Section D, you may assign appropriate activities from the Try This! section, located on pages 39–42 of the *Take a Chance* Student Book. The Try This! activities reinforce the key math concepts introduced in this section.

Planning Assessment

- Problem 2 can be used to informally assess students' ability to use repeated trials of a single experiment to estimate chances.
- Problem 10 can be used to informally assess students' ability to list the possible outcomes of simple chance and counting situations.
- Problem 12c can be used to informally assess students' understanding of the fact that variability is inherent in any probability situation.
- Problem 15 can be used to informally assess students' ability to find chances, in percents, fractions, or ratios, in simple situations; use repeated trials of a single experiment to estimate chances; and to use tree diagrams to represent one-, two-, and three-event situations.
- Problems 30 and 31 can be used to informally assess students' ability to use tree diagrams to represent simple one-, two-, and three-event situations and to model real-life situations involving probability.

D. LET ME COUNT THE WAYS

Families

There are many different types of families. Some families have one adult.

Some families have two adults, and some families have more.

Some families have children, and some do not.

1. a. Suppose you look at 20 families with two children. How many of these families do you think will have one boy and one girl?

   b. Other students in your class may not agree with your answer to part **a**. Write a short explanation to help convince them that your answer is correct. Drawing a diagram may be helpful.

girl

boy

2. You can **_simulate_** a study of two-child families by tossing two pennies. A head will represent a girl, and a tail will represent a boy. Toss the two pennies 20 times. See how many families with one boy and one girl you get. Was the result the same as your guess for problem **1a?**

1. a. Answers will vary. Accept any answer that is supported by logical reasoning in part **b.**

Possible answer: About 10 families will have one boy and one girl.

b. Sample answer explanation:

There are four different outcomes; two out of the four outcomes give one boy and one girl. So half of the families will probably have one boy and one girl.

First Child	Second Child
boy	girl
boy	boy
girl	boy
girl	girl

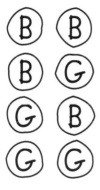

Sample incorrect answer explanation:

You can have one boy and one girl, two boys, or two girls. So, about one-third, or seven of the 20 families will have one boy and one girl.

1B and 1G

2B

2G

2. Answers will vary. Possible distribution of tosses:

two heads, five times; two tails, five times; one head and one tail, 10 times

Materials pennies (two per student or small group of students)

Overview Students list the possible combinations of boys and girls in two-child families and predict how many two-child families out of 20 they think would have one boy and one girl. After explaining their predictions, students simulate the same situation by tossing two coins.

About the Mathematics Efficient methods for recording the possible outcomes of a chance experiment are introduced in this section. Students use tables, organized lists, and visual models such as tree diagrams to find all possible outcomes. Students simulate real-life situations throughout this section, such as finding all possible combinations of shirts and pants to make different outfits.

Planning Students may work on problems **1** and **2** individually or in small groups. Be sure to discuss problem **2** as a class.

Comments about the Problems

1. a. Some students may incorrectly reason that $\frac{1}{3}$ of the families will have one boy and one girl (since there are three possible outcomes). However, there are actually four possible outcomes for two children: boy-boy, boy-girl, girl-boy, and girl-girl. Since boy-girl and girl-boy each result in a family with one boy and one girl, their combined chance is $\frac{2}{4}$ or $\frac{1}{2}$.

b. You might want to initially discuss this problem with students to help clarify their reasoning. It is not important that all students be convinced of any particular answer at this point, since this topic is revisited on page 34 of the Student Book.

2. Informal Assessment This problem assesses students' ability to use repeated trials of a single experiment to estimate chances. You may want to tabulate the class results to see if, indeed, half of the total tosses with two pennies resulted in a boy and a girl. A large number of coin tosses tends to reduce the variability of the outcomes. The pattern is then easier to see.

Robert's Clothes

Here are Robert's clothes.

This is the pair of pants and the T-shirt that Robert wears to school most often.

3. Find a way to show all of the outfits that Robert can wear to school. How many outfits are there?

4. Hillary bought Robert a new T-shirt when she went to the Compass Rose concert. How many different outfits can Robert wear now? Explain your answer.

5. How many outfits could Robert wear if he had four shirts and three pairs of pants?

3. There are six different outfits that Robert can wear.

pants	shirts
2	3

so there are 2 X 3 = 6 possibilities.

4. There are eight possible outfits. Explanations may vary. Sample explanation:

The new shirt can be worn with either pair of pants, making a total of two new outfits. Adding the two new outfits to the six outfits Robert already has makes a total of eight different outfits.

5. 12 different outfits

Materials copies or overhead transparency of the clothes shown on page 28 of the Student Book, optional (one copy per student or one transparency)

Overview Students find efficient ways to record the total number of possible outfits that can be made from three shirts and two pairs of pants, and from four shirts and two pairs of pants.

About the Mathematics The answers in the solutions column are presented in an informal tree diagram. Each pair of pants can be matched with three different shirts to make three combinations. Since there are two pairs of pants, there are a total of six combinations (2 × 3 = 6). When one more shirt is added, the number of combinations with each pair of pants becomes four, making a total of eight combinations (2 × 4 = 8). Multiplying the number of pants by the number of shirts to find the total possible combinations illustrates an important concept that is developed in this section. Using the visual model of the tree diagram will help students understand this property of multiplication.

Planning To prepare for problem **3,** you may want to make copies of the shirts and pants so that each student can cut them out. Or you can make an overhead transparency and have students find the combinations using the overhead projector. You may have students work on these problems individually or in small groups. Be sure to discuss problem **5** as a class.

Comments about the Problems

3. Students may use a variety of strategies:
- make drawings of all possible outfits,
- cut out the clothes from the copy and show all the combinations,
- record the combinations in a chart,
- make all the combinations on the overhead projector.

4. Some students may need repeat the entire process over again to find the total combinations. Other students may realize that adding one more T-shirt adds two more combinations, making a total of eight possible outfits.

5. Some students may be able to find the answer by generalizing from their results in problems **3** and **4:** for every pair of pants there are four possible T-shirts. Since there are three pairs of pants, the total number of possibilities is 12 (3 × 4 = 12).

Hillary's Clothes

One day, Hillary decided to choose a shirt and a pair of pants with her eyes closed.

6. What outfit would you expect to see Hillary wearing?

7. How could Hillary have used number cubes to help her choose her clothes?

Hillary says,

THE CHANCE THAT I WILL PICK MY FROG SHIRT WITH CHECKERED PANTS IS 1 OUT OF 36.

8. Is Hillary right? Explain.

6. Hillary will most likely wear a white shirt with white pants because there are more white shirts and white pants than anything else in her wardrobe.

7. Assign each of the six shirts to one of the six numbers on one number cube and each of the six pants to one of the six numbers on a different number cube. A single roll of both cubes will result in one outfit.

8. Yes. There are 36 different outfits she can wear. The frog shirt and the checkered pants together are one of the 36 outfits.

Overview Students predict which outfit a girl would choose from a wardrobe of six shirts and six pairs of pants. They also explain how she could use number cubes to help her make her choice.

About the Mathematics Tree diagrams (like those shown in the solution for problem **3** on page 67 of this Teacher Guide) can be helpful for finding the total number of possible combinations. A chart or table is another helpful method in which to record the possible combinations in a systematic way.

In problem **6,** since there are six pairs of pants and six T-shirts, there are 36 possible combinations. A table or tree diagram will show that six of these combinations are the same: a pair of white pants and a white T-shirt. This can be explained by reasoning that there are two white pants and three white T-shirts, making a total of six possible combinations. Since 6 out of 36 combinations are the same, the chance is 6 out of 36 (or one out of six) that there is a combination of a white T-shirt with white pants. Most students will not reason on this abstract level, but get their answer by reading the tree diagram. Others may find the answer without finding all the possible combinations.

Planning Students may do problems **6–8** in pairs or small groups. It is important that students are able to explain where the number 36 comes from. (There are a total of 36 possible outfits.)

Comments about the Problems

6. If students are having difficulty, encourage them to study the different shirts and pants pictured and discuss the characteristics of Hillary's clothes; or you may suggest that they skip this problem for now and return to it after they solve problem **8.**

8. Discuss students' strategies. Some students may need to draw a tree diagram to find that there are 36 combinations, while others may make a table or a list. Some may reason that for each pair of pants, there are six different T-shirts. There are six pairs of pants, making a total of 36 possibilities (some of which look the same). Since drawing is time consuming, some students may use shortcuts, with colors, letters, or numbers representing the different shirts and pants.

Extension When comparing different combinations of white pants and white shirts, you may want to discuss what is different and what is not different. [They look the same, but different items of clothing are involved.]

Open or Closed?

If you are with a group of three people, here is a way of choosing one person.

Stand in a circle, facing each other. One of you (or everyone at once) says: "One, two, three . . . go!"

At "go," each person puts out either an open hand or a closed fist.

Hillary, Robert, and Kevin played the game. Each winner is shown on the right in the table below.

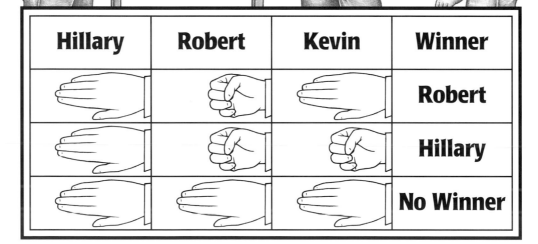

Hillary	Robert	Kevin	Winner
			Robert
			Hillary
			No Winner

9. Describe another situation in which there is no winner if you use this method.

10. How many combinations of open and closed hands are there in the game? List as many as you can.

11. Do you think this is a fair way to decide something? Why or why not?

9. Three closed fists will produce no winner, since the winner is the person whose hand does not match the others' hands.

10. There are eight possibilities:

Hillary	Robert	Kevin	Winner
open	open	open	none
open	open	closed	Kevin
open	closed	open	Robert
open	closed	closed	Hillary
closed	open	open	Hillary
closed	open	closed	Robert
closed	closed	open	Kevin
closed	closed	closed	none

11. Yes, this method is fair. There are a total of eight possible outcomes, and each person can win two ways—if his or her hand is open and the other two hands are closed, and if his or her hand is closed and the other two hands are open. So each person has a two out of eight (or one out of four) chance of winning.

o = open
c = closed

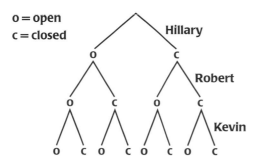

Overview In a game situation using open hands and closed fists, students determine all the possible outcomes and decide whether or not the game is fair.

About the Mathematics To decide whether or not this is a fair game, the number of favorable outcomes must be compared with the total number of possible outcomes.

Listing all possible combinations in a systematic way is important to ensure that no possible outcome is accidentally omitted. Students made organized lists of possible outcomes in the grade 5/6 unit *Patterns and Symbols*. When students look at all possible outcomes, it becomes clear that each of the three players has an equal chance of winning; that is, for each player, two out of the eight possible outcomes are favorable.

Planning You can have students play the game in small groups. Discuss problem **11** with the whole class to ensure that all students understand how the game works and that it is a fair game. You can use problem **10** for assessment.

Comments about the Problems

10. Informal Assessment This problem assesses students' ability to list the possible outcomes of simple chance and counting situations. This game is different from the pants and T-shirt situation, in which there are two choices. Students should realize that there are actually three situations here in which a choice has to be made: one for Kevin, one for Robert, and one for Hillary. Each person has two choices (open or closed fist), so there are eight total possibilities ($2 \times 2 \times 2 = 8$).

Since this game has two possible outcomes, with an equal chance for each person (open or closed), the game can be simulated with coins. Heads represents open and tails represents closed (or the other way around). Be aware that the game, as it is presented here, is only fair when played by three people.

11. Make sure students explain their reasoning in terms of chance and probability.

Extension Discuss with students how they could play the game with more than three people, and whether or not the game would still fair.

Mazes

Hillary has a pet mouse named Harry. She is bringing Harry to school today because the class is running mice through mazes for an experiment. The mice are put in at one end of the maze. At the other end, there is food in room 1, 2, 3, or 4. The mice cannot tell (or smell!) which door leads to the food. They, therefore, have an equal chance of going through any of the doors.

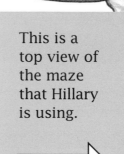

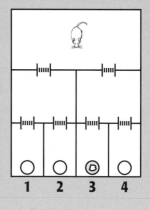

This is a top view of the maze that Hillary is using.

1 2 3 4

When the students bring all of the mice that they own to school, there are 60 in all.

12. a. Suppose you put 60 different mice in Hillary's maze, one after the other. About how many of these 60 mice would you expect to go through the first door on the right?

b. About how many would you expect to end up in room 3, where the food is?

c. Would exactly that many end up in room 3?

d. Is the chance that a mouse will end up in room 3 greater than, less than, or equal to the chance that it will end up in room 2? Why do you think so?

12. a. about 30 mice

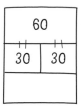

b. about 15 mice

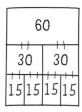

c. You cannot say for sure. The doors that each mouse will choose are unpredictable. Since there are only 60 trials (which is a relatively small number), the result would probably not be exactly 15 mice, but would be close.

d. The two chances are equal. There is a 50-50 chance that a mouse will choose either of the first two doors and a 50-50 chance that a mouse will choose either of the next two doors. Since the chances are equal for each room, 15 mice should end up in each.

Overview Students predict how many of 60 mice would choose one of four different rooms in a maze and explain the rationale behind their predictions.

About the Mathematics In previous problems, students had to find the total number of combinations. In these problems, the total number is given (four rooms). Students must predict the number of mice that will enter each of the four rooms. The assumption is made that the mice have no preference for any door, so at every intersection, each possible direction has the same chance of being chosen. It is not possible to actually do this experiment with 60 mice; however, you can simulate it (see Planning below).

On this page and the following, the transition is made from finding the expected value (how many of the *n* mice do you expect to end up in room one?) to finding the individual chance (what is the chance that a mouse will end up in room one?) This is a very difficult transition. Do not push students to express the chances using fractions such as $\frac{1}{3}$. Students can use ratio expressions like 10 out of 60 mice. The idea of individual chance is revisited extensively in later grade 8/9 units.

Planning You may want to make a copy of the maze on an overhead transparency so that students (or you yourself) can show how the mice split at each intersection. Theoretically, for every individual mouse, there is a 50% chance that the mouse will go right or left. You may, therefore, simulate the experiment by tossing a coin 60 times to determine whether a mouse will go right or left. Simulating the experiment using coins will probably not result in an equal distribution of the mice over the four rooms. You may have students do the problems on this page in pairs or small groups. Discuss problem **12d** in class.

Comments about the Problems

12. a. Have students write the number of mice that are expected to pass through each door in the maze. Refer to the text, where it is stated that the mice do not prefer any of the doors over the others.

   b. Have students write the number of mice that would end up in each room in the maze.

   c. Informal Assessment This problem assesses students' understanding of the fact that variability is inherent in any probability situation. Make clear that, contrary to **a** and **b,** this question has to do with the actual experiment. Although, theoretically, half the mice would go right and the other half left, this may not actually happen if you let mice run through the maze.

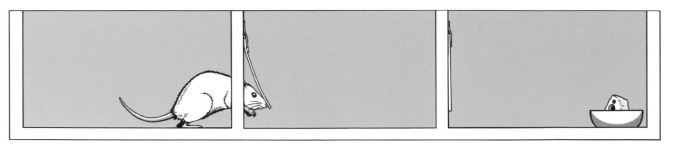

Each mouse has two different choices to make before reaching one of the final rooms. To describe the choices, you can use a picture called a **_tree diagram._**

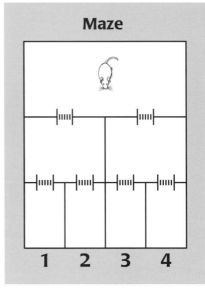

Maze

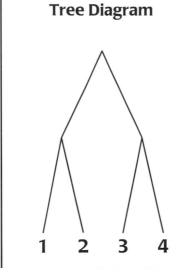

Tree Diagram

1 2 3 4

13. a. Use **Student Activity Sheet 4.** Put an H on the tree diagram at the place Harry would start.

b. If Harry finds the food in room 3, trace Harry's path on the maze.

c. Trace Harry's path on the tree diagram.

Robert wonders how the mice would behave in a different maze.

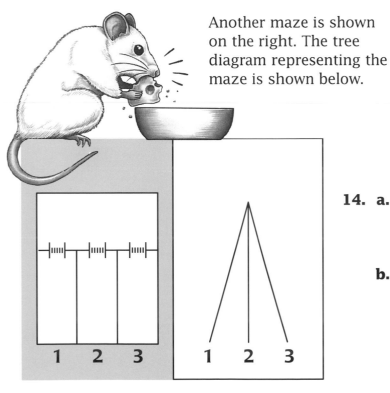

Another maze is shown on the right. The tree diagram representing the maze is shown below.

14. a. If you put 60 mice in this maze, about how many would you expect to end up in room 3?

b. What is the chance that a mouse would end up in room 3 for this maze?

13. a.

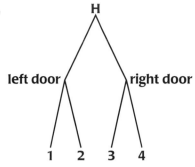

b.

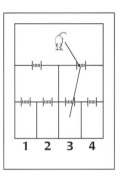

c.

14. a. About 20 mice would end up in room 3.

b.

```
          60
        /  |  \
      20  20  20
```

The chance that a mouse would end up in room 3 is 20 out of 60, 1 out of 3, or $\frac{1}{3}$. Accept answers expressed using a percent, a ratio, or a fraction.

Materials Student Activity Sheet 4 (one per student)

Planning The activity from the previous page of the Student Book continues as students trace the path of one mouse through different mazes by making tree diagrams.

About the Mathematics In the problems on this page, each maze is represented by a tree diagram. The number of mice that would travel in each direction is written next to that branch in the diagram. A more abstract method involves writing the chance (expressed as a fraction) next to each branch. In that way the tree diagram becomes a chance diagram, or chance tree. Writing whole numbers to represent mice is better at this stage. The transition to a chance diagram is made in grades 7/8 and 8/9.

You can indicate each path through a maze or a tree diagram by assigning each path or branch a different color. This may help students understand how to better read the tree diagram.

Planning Students may work on problems **13** and **14** in pairs or in small groups. Problem **14** may be assigned as homework. Check the results in the next class session.

Comments about the Problems

13. The route in the maze is the same as the path in the tree diagram. Students can draw the tree diagram inside the maze.

14. Homework This problem may be assigned as homework. For part **a,** in this maze there is only one intersection; in other words, there is only one point at which a choice has to made. Students can assign each branch the correct number of mice (20, again), assuming that the mice have no preference.

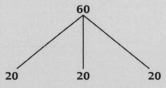

b. If students are having difficulty, ask them how many of the 60 mice they think will end up in room 3. You might want to ask this same question for other numbers of mice such as 30, 15, 12, 6, or even 3. Allow students to express their answers using ratios (so many mice out of the total number of mice) rather than expressing them as fractions.

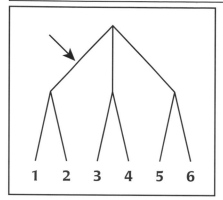

15. Suppose the class puts 60 different mice in a maze that goes with the tree diagram on the left.

  a. About how many of these 60 mice would you expect to take the path that the arrow points to?

  b. About how many would you expect to end up in room 2?

  c. What is the chance that one mouse starting at the beginning will end up in room 2?

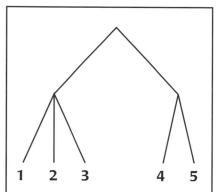

16. a. On the left is a tree diagram for a new maze. About how many of the 60 mice would you expect to end up in each room for this maze?

  b. What is the chance that a mouse would end up in room 3 for this maze?

17. On the left is a tree diagram for a different maze.

  a. What is the chance that a mouse will take the first door on the left? the middle door?

  b. What is the chance that a mouse will end up in room 2? room 3? room 6?

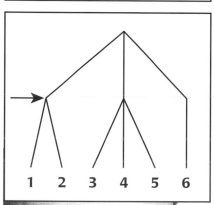

15. a. about 20

   b. about 10

   c. 10 out of 60, 1 out of 6, or $\frac{1}{6}$.

16. a. 10 mice in rooms 1, 2, and 3; 15 mice in rooms 4 and 5.

   b. 10 out of 60, 1 out of 6, or $\frac{1}{6}$.

17. a. First door on the left: $\frac{1}{3}$, 30 out of 90, or 1 out of 3

     Middle door: $\frac{1}{3}$, 30 out of 90, or 1 out of 3

   b. Room 2: $\frac{1}{6}$, or 1 out of 6

     Room 3: $\frac{1}{9}$, or 1 out of 9

     Room 6: $\frac{1}{3}$, or one out of three

90 mice:

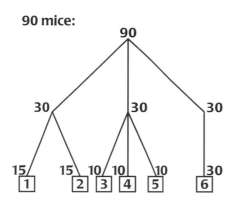

Overview Students again use tree diagrams to predict in which rooms 60 mice would end up in three different mazes.

About the Mathematics As previously stated, students' expectations may differ from the results of an actual experiment using coins. It is important to remind students of the fact that expectations are not always realized.

In the second and third mazes on this page, the chances for each room are no longer equal. As on the two previous pages, the chance of ending up in a room is actually a proportion of the total number of mice. Therefore, allow students to express these chances as ratios. Some students may express chances using fractions. They may reason that from the start, $\frac{1}{2}$ go to the left; then $\frac{1}{3}$ of this half goes to room two, so $\frac{1}{6}$ of the mice end up in room two. Do not require all students to think this abstractly.

Planning Problems **15–17** may be done in pairs or in small groups. You may use problem **15** as an assessment and problem **16** as homework. Be sure to discuss problem **17** in class to ensure that students understand how to use the tree diagram to calculate the number of mice that will end up in each room.

Comments about the Problems

15. Informal Assessment This problem can be used to assess students' ability to find chances, in percents, fractions, or ratios, in simple situations; use repeated trials of a single experiment to estimate chances; and to use tree diagrams to represent one-, two-, and three-event situations.

If students are having difficulty, you may have them write down the number of mice that follow each path next to each branch in the tree diagram on their own paper.

16. Homework This problem may be assigned as homework. Make sure that students understand why an equal number of mice do not end up in each room.

17. If students are having difficulty, point out that they can first choose a number of mice and imagine those mice running through the maze. Note that if they choose 60 mice, they may have difficulty deciding how the 20 mice that pass through the middle door divide themselves between rooms 3, 4, and 5. You may suggest an easy number, such as 90, to use here.

Two Children Again

Tree diagrams can be useful for more than just mice.

Consider families with children once more.

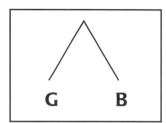

A tree diagram can show the two possibilities for one child.

18. Each path on this tree diagram has an equal chance. What is the chance of having a girl?

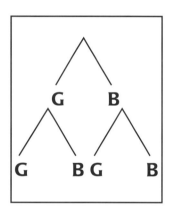

If a family has a second child, you can extend the tree diagram like the one on the left.

19. a. Copy the tree diagram and trace the path for a family that had a girl first and then a boy.

    b. What are all the possible combinations for a family with two children?

    c. If there are 20 families with two children in Hillary's class, how many would you expect to have two girls? What is the chance that a family will have two girls?

    d. Is the chance greater to have two girls or to have a boy and a girl? Explain.

Now look at families that have three children.

20. a. Extend the tree diagram to show a third child.

    b. List all of the different possibilities for a family with three children.

    c. Robert says, "It's less likely for a family to have three girls than to have two girls and a boy." Explain this statement.

    d. Make some other statements using the tree diagram from part **a.**

18. one out of two, $\frac{1}{2}$, or 50%

19. a.

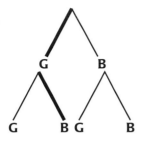

b. BG, GB, BB, and GG

c. about five, which is equivalent to one out of four, or $\frac{1}{4}$ of the families

d. A family has a better chance of having a boy and a girl than two girls, since the chance of having a boy and a girl is $\frac{1}{2}$, and the chance of having two girls is $\frac{1}{4}$. One-half is greater than one-fourth.

20. a.

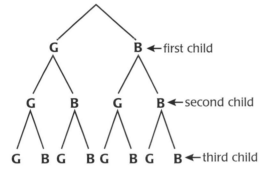

b.

BBB	GBB
BBG	GBG
BGB	GGB
BGG	GGG

c. There is only a $\frac{1}{8}$ or one out of eight chance of getting three girls. There is a $\frac{3}{8}$ or three out of eight chance of getting two girls and one boy.

d. Answers will vary. Sample responses:

• The chance of having two boys and one girl is the same as the chance of having two girls and one boy.

• The chance of having three boys is the same as the chance of having three girls.

• The chance of having three boys is less than the chance of having one boy and two girls.

Overview Students use tree diagrams to investigate the possible combinations of boys and girls in one-, two-, and three children families.

About the Mathematics Tree diagrams are again used to represent all possible outcomes. These diagrams cannot be interpreted in exactly the same way that the maze diagrams were. To find all possible outcomes with the mazes, one had only to look at the endpoints. To find all possible combinations of boys and girls from these tree diagrams, one has to carefully trace each path. At each intersection, a boy or a girl is chosen. Different paths can lead to the same combination of boys and girls, but in a different order.

Planning Students may work on these problems in pairs or in small groups. After problem **19b,** discuss how students can find all the combinations from the tree diagram by tracing each path and writing down each letter they pass. You may also want to discuss situations in which the order of the letters in each path is important and situations in which it is not. For example, in problem **19a,** the order is important, while in problems **19c** and **19d,** it is not. Problem **20** can be assigned as homework. List students' responses for problem **20d** on the board and discuss each one.

Comments about the Problems

18. Allow students to express chances in different ways: in words, ratios, fractions, and percents.

19. Make sure students know how to read the diagram. Compare their answers to the answers for the problems on page 27 of the Student Book.

> **b.** The term *combinations* refers to the birth orders for one boy and one girl.
>
> **c.** Refer to the strategy used with the mice in the mazes. Have students label each branch with a number or with an endpoint.
>
> **d.** Make sure students understand that the order does not matter.

20. Homework This problem may be assigned as homework. You may want to point out that in the tree diagram, each row or line corresponds to having one more child.

Writing Opportunity Have students write a letter to someone in which they explain one or two statements from their answer to Problem **20d.**

Extension You may have students investigate the possible combinations in four-child families.

Activity

Sum It Up

21. Roll two different-colored number cubes. For each pair of numbers that can come up, what are the different sums you can get?

Hillary and Robert sometimes play Sum It Up during lunch.

They each pick one of the possible sums of two number cubes. Then they roll the number cubes, and the first person to roll his or her sum four times wins. The loser has to clean up the other's lunch table.

22. Which sum do you think would be best to pick?

Play the game with the person next to you. If you both want the same number, come up with a fair way to decide who gets the number.

23. Record the sums that are thrown. What was the winning sum in your game?

Play the game five or six times. You can change numbers if you want.

24. Now what do you think is the best sum to choose?

21. You can roll sums from 2 to 12.

22. Answers will vary. Actually, seven is the best choice. Seven has the highest probability of being rolled because there are the greatest number of ways to roll a sum of seven.

23. Answers will vary.

24. Seven is the best sum to choose.

Materials colored number cubes (two different-colored cubes per pair of students)

Overview Students first determine all possible sums that can be obtained when tossing two number cubes. They then play a game with a partner involving the two cubes. Each player first chooses one of the possible sums that he or she thinks will occur most often. The players then take turns tossing the number cubes four times and recording the sum of each toss. The player whose chosen sum occurs more often is the winner. After playing the game five or six times, students decide which of the possible sums is the "best" sum to choose when tossing two number cubes.

About the Mathematics This complex activity is similar to those in Section B. Students again investigate the differences between experimental and theoretical chances. Since students have had more practice with listing the possible outcomes of multi-event situations, they can now understand the theory behind predicting the outcomes of rolling two number cubes. On the next page, they investigate a two-event situation using a chart and a tree diagram.

Planning Have students do this activity in pairs or in small groups. Ask them to list their outcomes. While discussing the answers to problem **24,** watch to see if all students come up with the same pattern in the outcomes. You do not need to discuss explanations or go into the methods for finding them. Students will do that on the next page.

Comments about the Problems

22. At this point, many students may have no idea which sum is best to pick. If students are having difficulty, encourage them to make a guess.

23. Initially, some pairs of students may each pick a number that does not come up often. If you have them record all the sums that come up during the game, they may develop a feeling for sums that come up more often. During the next game, they will be more apt to choose one of these numbers. Students are not required to list all possible outcomes here. This will be done on the next page.

24. Although seven is the best sum to pick, since it has the highest probability of being rolled, accept any answer.

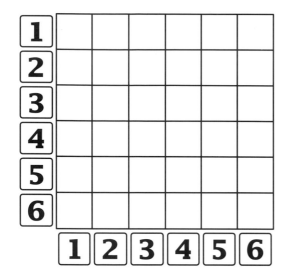

25. Student Activity Sheet 5 has a grid showing the possible numbers for each of two number cubes.

   a. For each square, fill in the sum of the numbers.

   b. How many different combinations are possible when you roll two number cubes?

   c. How many ways can you get a sum of 10 with two number cubes?

   d. What is the best number to pick if you are playing Sum It Up? Is your answer different from your choice in problem **24?**

26. a. Draw a tree diagram to show all of the possible combinations for rolling two number cubes. It might be messy!

   b. Color the squares in the grid from problem **25** and the paths in the tree diagram from part **a** that give a sum of 10.

27. a. What is the chance of rolling two 1s? What is the chance of rolling doubles?

   b. What is the chance of rolling a 7?

28. What do you think is the chance of *not* getting a 10? *not* getting a 7?

25. a.

1	2	3	4	5	6	7
2	3	4	5	6	7	8
3	4	5	6	7	8	9
4	5	6	7	8	9	10
5	6	7	8	9	10	11
6	7	8	9	10	11	12

1 2 3 4 5 6

b. 36 combinations

c. A sum of 10 can be thrown three different ways: 6 + 4, 4 + 6, and 5 + 5.

d. Seven. Answers will vary, depending on the sum students chose in problem **24.**

26. a.

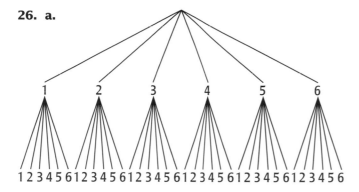

b.

1	2	3	4	5	6	7
2	3	4	5	6	7	8
3	4	5	6	7	8	9
4	5	6	7	8	9	10
5	6	7	8	9	10	11
6	7	8	9	10	11	12

1 2 3 4 5 6

27. a. The chance of rolling two ones is 1 out of 36. The chance of rolling doubles is 6 out of 36 or one out of six.

b. The chance of rolling a 7 is 6 out of 36.

28. Since there are three ways of getting a 10, there are 33 way of not getting a 10. So there is a 33 out of 36 chance of not getting a 10. Similarly, since there are six ways of getting a 7, there are 30 ways of not getting a 7. So there is a 30 out of 36 chance of not getting a 7.

Materials Student Activity Sheet 5 (one per student)

Overview Students fill in charts and draw tree diagrams to show all the possible combinations for rolling two number cubes. They then investigate the chances of different outcomes.

About the Mathematics Here the theory behind the game Sum It Up is investigated. Students discover that some sums occur in more ways than other sums. If students have difficulty distinguishing 4 + 6 from 6 + 4, you may want to assign different colors to the number cubes. For example, 4 blue + 6 red is not the same as 6 blue + 4 red. You can also label the table with the two color names. Counting all the possibilities in the table is also an option.

Planning Students may work on these problems in pairs or small groups. You may assign this whole page as homework or have students solve problem **25** in class, discuss the results, and assign the rest of the problems as homework. Be sure to discuss problem **28** on the following day. If you find pages 35 and 36 take too much time, you may have students play the game less often and skip problem **26.**

Comments about the Problems

25. You might discuss how the different sums from the grid come from the number cubes. Here it may be helpful to use colors. For example, the 7 in the top right corner of the table may be formed by a red 6 and a blue 1, whereas the 7 in the lower right corner of the table would have the two colors switched. By comparing the answer for problem **24** with that for **25d**, you can call students' attention to the difference between theoretical and experimental chance, as explored in Section C.

26–28. Homework These problems may be assigned as homework.

26. Ask students to explain how the total number of outcomes and the different sums can be located in the tree diagram.

27. Chances can be calculated by comparing the number of favorable outcomes to all possible outcomes.

28. You may give students the option of coloring the favorable sums in the chart for these two special cases.

Treasure

During a hike along the shoreline at low tide, Hillary and Robert find a big chessboard on the sand.

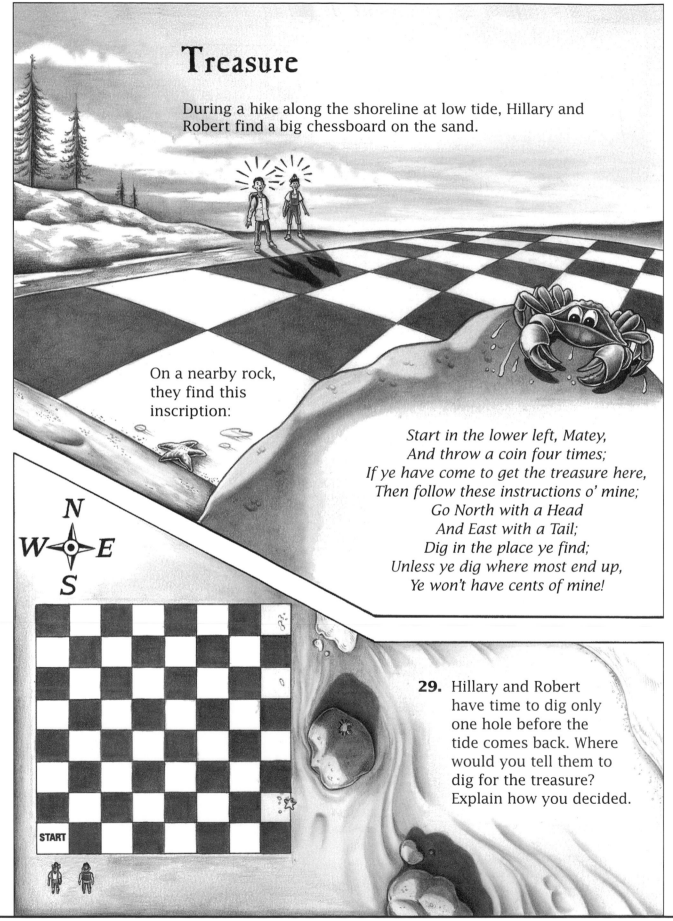

On a nearby rock, they find this inscription:

*Start in the lower left, Matey,
And throw a coin four times;
If ye have come to get the treasure here,
Then follow these instructions o' mine;
Go North with a Head
And East with a Tail;
Dig in the place ye find;
Unless ye dig where most end up,
Ye won't have cents of mine!*

29. Hillary and Robert have time to dig only one hole before the tide comes back. Where would you tell them to dig for the treasure? Explain how you decided.

29. The most common outcome of throwing a coin four times is two heads and two tails. Therefore, the students should go north twice and east twice and dig there.

Other ways to solve this are to list possible directions:
EEEE,
EEEN,
and so forth, and then draw a tree diagram, as shown below.

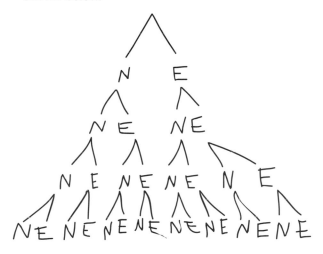

There are six possible paths in which a person moves north and east two each.

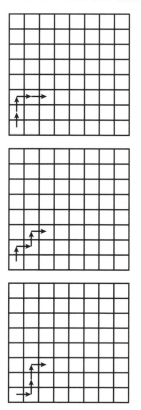
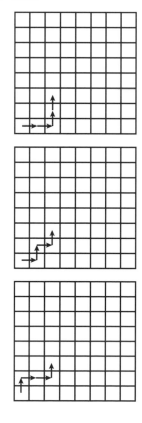

Materials copies of the chessboard on page 84 of the Teacher Guide, optional (one per pair or group of students)

Overview Students help Hillary and Robert plan a route to dig for buried treasure on a large chessboard. Students first read a riddle that instructs them to:
• begin the game in the square in the lower left-hand corner of the chessboard,
• throw a coin four times, traveling north if the coin lands on heads and east if the coin lands on tails,
• dig in the final location.

In other words, the four coin tosses must be one of the possible outcomes that occur most frequently when tossing four coins (two heads and two tails). Students then draw a route that they hope will lead to buried treasure (see About the Mathematics below).

About the Mathematics There are 16 possible outcomes when tossing a coin four times. Since each toss can be a head or a tail, there are $2 \times 2 \times 2 \times 2 = 16$ outcomes. Of these, six will result in two heads and two tails: HHTT, HTHT, HTTH, TTHH, THTH, and THHT. This is the most likely outcome. Therefore, most students will tell Hillary and Robert to dig at a spot that is located two squares east and two squares north from the starting point.

Planning Students may work on problem **29** in pairs or in small groups. You may want to make copies of the chessboard for each pair or group of students so that they can draw on it if they wish. You might also bring in a real chessboard and coins so that students can model the activity. This problem is optional.

Comments about the Problems

29. A tree diagram is probably the easiest way to see that two heads and two tails are the most likely outcome. Students may put a tree diagram right on the chessboard; the square with the most paths to it is the place in which to dig.

Summary

In this section, you learned that counting the number of ways an event can occur can help you find the chances of the event.

You can write all of the possible ways that something can occur, or you can draw pictures. Tree diagrams are one type of helpful picture.

A tree diagram can give information about:

• all possible outcomes,

• the chance that any single outcome will occur.

Here is a tree diagram of problem **3** (Robert's clothes) from page 28.

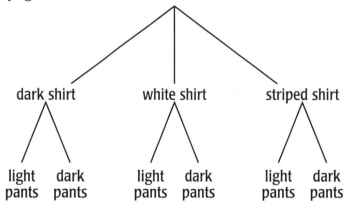

Summary Questions

30. If Robert picks his clothes at random, what is the chance that he will pick a striped shirt and light pants?

31. If a tree diagram ends with eight branches, is the chance of each outcome the same? Give an example to support your answer.

30. The chance is $\frac{1}{6}$ or one out of six.

31. Not necessarily. Here are two tree diagrams, the first showing an equal chance for each outcome and the second showing different chances for the outcomes.

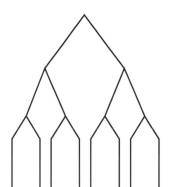 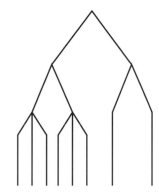

Overview After reading the Summary, which reviews the main topics covered in this section, students solve two problems about chance and tree diagrams.

Planning You can use the summary questions to assess students' understanding of tree diagrams and chances. After students complete Section D, you may assign appropriate activities from the *Try This!* section, located on pages 39–42 of the *Take a Chance* Student Book, for homework.

Comments about the Problems

30–31. Informal Assessment These problems assess students' ability to use tree diagrams to represent simple one-, two-, and three-event situations and to model real-life situations involving probability.

30. Allow students to express their answers in different ways.

31. If this problem is too abstract for students, refer to the mazes on pages 31–33 of the Student Book. Have students reformulate the problem in terms of the mazes or present the reformulated problem to students, for instance, by asking, *Is it possible to design a maze with eight rooms so that a different number of mice end up in each room?* [Yes, see solutions.] Then have students draw a maze before answering the question.

Extension You can ask other questions using the tree diagram of Robert's clothes. You can also have students add additional clothing items to the tree diagram and formulate and answer their own questions.

Assessment Overview

Students work on six assessment problems that can be used at the end of the unit. You can evaluate these assessments to determine what each student knows about chance and probability, and which strategies they use to solve each problem.

Goals

- determine whether or not a simple experiment is fair

- describe chance in everyday language

- estimate chances in percents, from 0% to 100%

- find chances, in percents, fractions, or ratios, for simple situations

- list the possible outcomes of simple chance and counting situations

- understand that in independent trials, the next outcome is not affected by previous outcomes

- use repeated trials of a single experiment to estimate chances

- use tree diagrams to represent simple one-, two-, and three-event situations

- understand that variability is inherent in any probability situation

- model real-life situations involving probability

- develop an understanding of the difference between theoretical and experimental probability

Assessment Opportunities

Problems 3a and 3c

Problems 1 and 3d

Problems 2 and 6b

Problems 2, 3d, 4b, and 4c

Problem 3b

Problem 5

Problem 4b

Problems 4a, 4b, and 4c

Problem 5

Problems 3c and 6a

Problems 3a and 5

Pacing

- The first five assessment problems will take approximately one or two 45-minute class session. The sixth problem may be assigned as homework. For more information on how to use the six problems, see the Planning Assessment section on the next page.

About the Mathematics

These six end-of-unit activities assess the major goals of the *Take a Chance* unit. Refer to the Goals and Assessment Opportunities section on the previous page for information regarding the goals that are assessed in each activity. The problems do not ask students to use a specific strategy (express chance as a fraction, a ratio, a percent, or in words). Students have the option of using their own strategies or any of strategies with which they feel comfortable. They may also use any of the models that are introduced and developed in this unit (tree diagrams, chance ladders, or a circular model like the spinner). Students are not expected to use any formal algorithms to solve these problems. Students should demonstrate understanding and not mastery of dealing with chance and probability.

Materials

- Assessments, pages 106–109 of this Teacher Guide (one of each per student)

Planning Assessment

You may decide to let students work on these assessment problems individually if you want to evaluate each students' understanding and abilities, or you may have the students complete the first five problems in pairs or in small groups and do the sixth problem individually. Make sure you allow enough time for students to complete the problems. Students are free to solve the problems in their own ways. They may use any of the models to solve problems that do not ask for a specific model.

Scoring

In scoring the assessment problems, the emphasis should be on the strategies used rather than on students' final answers. The strategy a student chooses may indicate how well he or she understands the concepts of chance and probability. For example, a concrete strategy supported by drawings may indicate a deeper understanding than an abstract, computational answer. Consider how well the students' strategies address the problem as well as how successful the students are at applying their strategies in the problem-solving process.

PROBLEMS ABOUT CHANCE

Use additional paper as needed.

1. The weatherman announced,

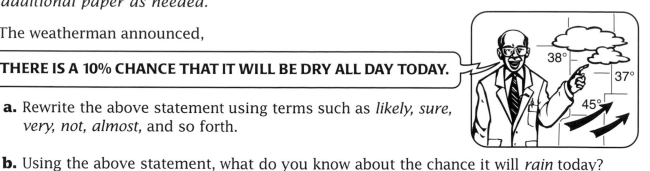

THERE IS A 10% CHANCE THAT IT WILL BE DRY ALL DAY TODAY.

   a. Rewrite the above statement using terms such as *likely, sure, very, not, almost,* and so forth.

   b. Using the above statement, what do you know about the chance it will *rain* today?

2. Next to each story below, write the number of the chance statement that matches it. Some stories may have more than one matching statement, and some statements may match several stories.

Story		**Statement**

  a. My friend raffled off a prize at a party. He made 15 tickets and gave me three. What are my chances of winning the prize? _____

               1. There is a 20% chance that this will happen.

               2. There is about a 30% chance.

  b. What is the chance that a frog will land on a black square on the floor on the left? _____

               3. It is not very likely.

               4. There is an 80% chance.

  c. I am playing a game. If I throw a 3 or a 5 with a number cube, I will win. Is it likely that I will win? _____

               5. The chances are 1 out of 3.

               6. There is a 50-50 chance.

  d.  Sam is thinking of a number between 1 and 15. What are my chances of guessing his number? _____

               7. The chances are 1 out of 5.

               8. The chance is 0%.

  e.  I throw two coins. What is the chance that I will get a head and a tail? _____

1. a. Answers will vary. Sample responses:

- It is very unlikely that it will be dry all day today.

- It is not very likely that it will be dry all day today.

- I am almost sure that it will not be dry all day today.

b. Answers will vary. Sample responses:

- There is a 90% chance that it will rain today.

- It is very likely that it will rain today.

- I am almost sure that it will rain today.

2. Possible matches are:

Story **a:** Statements 1 or 7; students can match statements 2 and 3 with Story **a** as long as they include an explanation.

Story **b:** statements 2 or 5

Story **c:** statements 2 or 5

Story **d:** statement 3

Story **e:** statement 6

Materials Problems about Chance assessment, page 106 of the Teacher Guide (one per student)

Overview Students interpret a chance statement, rewrite the statement, and find the complementary chance. They match stories with chance statements that reflect the likelihood of the events occurring.

About the Mathematics Problem **1** focuses on expressing chance in everyday language and as a percent. Translating from one expression into another is an important skill in critically analyzing a chance situation.

Chance statements can also be represented by drawings, such as tree diagrams and chance ladders. A visualization of a chance situation or statement may be helpful in determining whether or not a statement and situation match. Problem **1b** assesses students' understanding of the fact that the sum of all chances in any situation equals 100%, or one.

Planning You may let students complete these problems in pairs or in small groups. Students may also work individually.

Comments about the Problems

1. a. Students have the option of making drawings to visualize the situation. These drawings can be used as an intermediary step before students rewrite the statement using everyday language.

b. Students' responses should show their understanding of the fact that the chance that it will rain and the chance that it will be dry equals 100% or one.

2. Different strategies may be used to solve this problem. Some students may eliminate the possibilities that they know do not match before proceeding to match the stories with statements. Others may start with the chance story and find a matching statement, or vice versa.

Students can make drawings of each story and of each statement and then match the stories with the statements. Students may also rewrite the stories and the statements (in terms of likelihood, using percents, fractions, or ratios) and then match the statements with the stories.

PROBLEMS ABOUT CHANCE

Use additional paper as needed.

3. Katrina is a student in a small school that has only six classes. Two classes at her school are invited to visit a television studio in a nearby city. The principal wants to use a fair method to select which two classes will go.

    a. Describe two fair methods the principal can use to make the choice. Why do you think your methods are fair?

Mr. Harris's and Ms. Lyne's classes were picked. Katrina is in Ms. Lyne's class. She wonders what to wear. She takes all of her blouses and slacks out of the closet to see what outfits she might wear. She finds that she has 12 different combinations of a blouse and a pair of slacks.

    b. Describe or draw the clothes Katrina might have taken out of her closet.

In the television studio, four of the students will be chosen to be on camera. Mr. Harris's class has 23 students, 8 boys and 15 girls. Ms. Lyne's class has 27 students, 12 boys and 15 girls.

    c. Describe how you would pick the four students. Explain why you would use your method.

    d. Katrina is in Ms. Lyne's class. She wonders how good her chances are for being chosen. Using your method from part **c,** write her a note describing her chances.

3. a. Answers will vary. Sample responses:

- Put one number that represents each class in a bag and draw twice from the bag.

- Using one number cube, assign one number to each class in the school. Roll the number cube twice. If the same number comes up twice, roll the number cube a third time.

- Using two number cubes, assign one number to each class in the school. Roll the number cubes once. If a double number is thrown, the roll does not count. Roll again.

- Make a spinner with six equal-sized sections and spin it twice. If the same class is chosen again, spin a third time.

- Write all 15 combinations of two classes on individual cards. Put all the cards in a bag and choose one card.

These methods are fair because each of the six classrooms has the same chance of being chosen.

b. Answers will vary. Sample responses:
- four blouses and three pairs of slacks
- three blouses and four pairs of slacks
- six blouses and two pairs of slacks
- two blouses and six pairs of slacks
- twelve blouses and one pair of slacks
- one blouse and twelve pairs of slacks

c–d. Answers will vary. Sample responses:

- Select two students at random from each class. In this way, each class is equally represented, although students in Mr. Harris's class have a slightly larger chance of being chosen. If this method is used, Katrina will have a 2 out of 27 chance of being selected. Other students may describe this by saying *a small chance.*

- Choose one boy and one girl from each class. This would be a fair method if each class had an equal number of boys and girls, but this is not the case here. If this method is used, Katrina will have a 1 out of 15 chance of being selected.

- Group all the classes together and choose four students. Using this method, all four chosen students may be from the same class. Ignoring that fact, the method is fair because each student still has an equal chance of being chosen. Using this method, Katrina will have a 4 out of 50 chance of being chosen.

- Divide the boys and girls into two separate groups. Using this method, the boys would have a 2 out of 20 chance of being chosen. Katrina (and any other girl) would have a 2 out of 30 chance of being chosen.

Materials Problems about Chance assessment, page 107 of the Teacher Guide (one per student)

Overview Students design and describe fair selection methods. For one method, they find the chance of a particular outcome. In another situation, students find all possible combinations of blouses and pants given the total number of outfits.

About the Mathematics Problem **3a** assesses students' understanding of what a fair method is and whether or not they can create their own methods to make a fair decision. Problem **3b** asks students to find all possible combinations of blouses and pants when the total number of outfits is given. The last question deals with finding the chances for a given outcome of a simple experiment.

Planning You may let students complete these problems in pairs or in small groups. Students may also work individually.

Comments about the Problems

3. a. Be sure that students' representations (drawings, stories, or models) are used to describe two *different* methods of making a fair decision.

b. This problem is the reverse of those presented previously in the unit. Students are given the total number of possible combinations and must find combinations of blouses and pants that correspond to that total. They should recognize that there are several ways to arrive at 12 possible outfits.

c. Students may not think there is a reason to distinguish between boys and girls in this problem. It many instances, it does not make a difference, but it is often important to have both boys and girls represented in a sample.

d. Students' answers will depend on the method they designed in **3c.** As long as the reasoning used to answer **3d** is in line with their answer to **3c,** students should receive full credit.

Writing Opportunity You might ask students to write their responses for problem **3d** in their journals.

PROBLEMS ABOUT CHANCE

Use additional paper as needed.

4. a. Design two different mazes, each with eight rooms, so that each room has an equal chance of being reached.

   b. Four different tree diagrams of mazes are shown below. Order the diagrams according to the chance a mouse has of ending up in room 3. Describe how you determined the order.

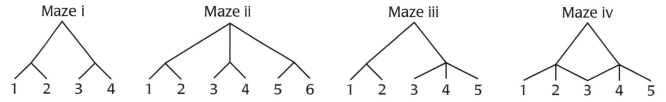

   c. Design a maze with eight rooms (different from those you created in part **a**) so that the chance of reaching room 5 is greater than the chance of reaching room 1. Find the chance of reaching room 1 and the chance of reaching room 5 in this maze.

5. Belinda found an old coin. She tossed the coin and heads came up on the first three tosses. What can you say about the coin?

4. a. Answers will vary. Some students may sketch something other than tree diagrams.

Sample tree diagrams:

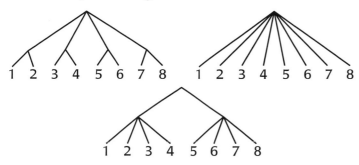

b. Accept chances expressed as fractions, ratios, or percents. Correct order of mazes from largest chance to smallest chance:

Maze iv ($\frac{1}{3}$ chance), Maze i ($\frac{1}{4}$ chance),

Maze ii ($\frac{1}{6}$ chance) or Maze iii ($\frac{1}{6}$ chance)

Explanations will vary. Sample explanation:

I found the chance of a mouse's entering room 3 in each maze and expressed each chance as a fraction. Then I ordered the fractions from greatest to smallest.

c. Answers will vary. Some students may sketch something other than a tree diagram. Students may express chances as fractions, ratios, or percents. Each chance answer should match its corresponding tree diagram. Sample responses:

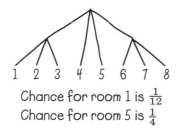

Chance for room 1 is $\frac{1}{12}$
Chance for room 5 is $\frac{1}{4}$

5. Answers will vary. Student responses should include the fact that three tries are not enough to decide whether or not a coin is fair. You need to conduct many more trials in order to find a pattern in the outcomes.

Materials Problems about Chance assessment, page 108 of the Teacher Guide (one per student)

Overview Students design a maze with rooms so that each room has an equal chance of being reached. They then design a maze so that each room has an unequal chance of being reached. Students interpret tree diagrams to find chances.

About the Mathematics A tree diagram is almost identical to a maze. The tree diagram shows the possible paths, and the intersections where choices must be made. To design mazes in which each room has an equal chance of being reached, the number of routes to each room must also be equal. Efficient counting methods and organized record keeping are necessary, especially in dealing with mazes that have several possible routes.

In any coin toss, the chances of landing on heads is always $\frac{1}{2}$. Although in a small number of trials, heads may not occur in exactly one-half of the trials. After many trials, the percent of heads will approach 50%. This dual aspect of chance is difficult to understand, and students are not expected to have grasped this completely yet.

Planning You may let students complete these problems in pairs or in small groups. Students may also work individually.

Comments about the Problems

4. a. The maze should have one entrance and eight rooms. The number of possible routes from the entrance to each room should be equal for each room.

b. This is a difficult question. Look at students' strategies to see whether they just counted the number of rooms, or if they ran mice through the maze. If students look only at the number of rooms, they will have the incorrect answer: i, iii and iv, ii.

c. Students may alter one of the mazes they used in problem **a.** They may also use the "counting the number of routes or paths" strategy to find the chances. Be sure students explain how they found their answers.

5. Students may use a tree diagram or another visualization to explain their reasoning for this problem.

PROBLEMS ABOUT CHANCE

Use additional paper as needed.

6. Paula and Huong made a board game for two players.

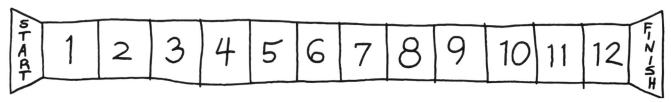

A sketch of their board is above. Below are the rules for the game which describe the four possible moves:

Move i: Go one step forward.
Move ii: Go three steps forward.
Move iii: Go three steps backward.
Move iv: Stay where you are.

They also decided to make some rules about *chance*. There must be a big chance for move i, equal chances for moves ii and iii, and a 25% chance for move iv.

They make a spinner to decide what move to make on each turn. The person who reaches Finish first is the winner.

a. Design a spinner that fits the rules Paula and Huong made.

b. Estimate the chance for each of the four moves, either as a fraction or as a percent.

After playing the game, Paula and Huong thought it took too long for the game to end. They decided to change the rules.

c. How would you change the rules to make the game go faster? Explain your answer.

6. a–b. Answers will vary. Two sections of the spinner must be equal in size to represent Moves ii and iii. One section should be larger than the other three sections to represent Move i. Move iv should be represented by a section that is exactly one-quarter of the spinner. Students can express the chances using percents, ratios, or fractions. When students use fractions, the chances should add up to one. When using percents, the chances should total 100%.

Sample responses:

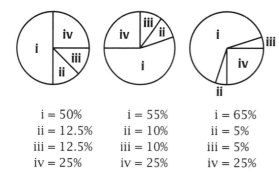

i = 50%	i = 55%	i = 65%
ii = 12.5%	ii = 10%	ii = 5%
iii = 12.5%	iii = 10%	iii = 5%
iv = 25%	iv = 25%	iv = 25%

c. Answers will vary, but should include an explanation. Sample responses:

- Move i can be changed to "Go four steps forward." Since this move has a greater chance of being selected on the spinner, the game should move along faster.

- You can assign a greater chance for Move ii and a smaller chance for Move iii by making the section of the spinner representing Move ii larger and by making the section of the spinner for Move iii smaller. In this way, there is a greater chance of moving three steps ahead than of moving three steps backward, and the game should go faster.

Materials Problems about Chance assessment, page 109 of the Teacher Guide (one per student)

Overview Students interpret a game and its rules and design a spinner that can be used to play the game. They also estimate chances for possible outcomes of the game and then change the rules to make the game move along faster.

About the Mathematics Interpreting the rules of a game just by reading the text may not be sufficient for all students. Some may need to play the game to fully understand the rules. Students estimate chances and express the chances using percents or fractions. Question **4c** assesses students' creativity.

Planning You may want students to work on this problem individually. This problem may be assigned as homework.

Comments about the Problems

6. a. The exact chance is given only for Move iv. Students must find the chances for Moves i, ii, and iii themselves. Any spinner will work as along as it matches the restrictions described in the problem.

b. In problem **6a,** students were not restricted to "easy numbers" that can be expressed as simple fractions or benchmark percents. Estimates should match the spinner they designed for problem **6a.**

c. Students may come up with a variety of answers, but they should keep in mind that it should still be possible to play the game. Students' explanations should be written in such a way that anyone who wants to play the game will understand the new rules. Students can opt to alter the chances on the spinner, the moves, or both.

Take a Chance
Glossary

The Glossary defines all vocabulary words listed on the Section Opener pages. It includes the mathematical terms that may be new to students, as well as words having to do with the contexts introduced in the unit. (Note: The student book has no glossary in order to allow students to construct their own definitions, based on their personal experiences with the unit activities.)

The definitions below are specific to the use of the terms in this unit. The page numbers given are from this Teacher Guide.

chance (p. 6) the possibility that an event will occur

fair (p. 8) giving equal chances to all possible outcomes

simulation (p. 62) a model of a process

tree diagram (p. 74) a picture with branches representing all possible choices or combinations
of choices

trial (p. 60) one of many repetitions of an experiment

Blackline
Masters

Dear Family,

Very soon your child will begin the *Mathematics in Context* unit *Take a Chance*. Below is a letter to your child that opens the unit, describing the unit and its goals.

You can help your child relate the classwork to his or her own life by asking for help in making fair decisions at home, such as choosing family members to set the table, empty the dishwasher, or clean a room.

Look for opportunities to bring chance and probability into the home. Discuss situations involving chance: the weather report, getting tickets for a special concert, or winning the lottery.

Have your child count the possible arrangements of the family members around the dinner table, in a church pew, or at a movie theater. Talk about how likely it is that any two people will sit next to each other. If you have games that use number cubes or spinners, discuss how chance is involved in each game. Do the games seem to be fair? Check newspapers for statements about chance and read them to your child.

Chance is an important concept in dealing with uncertainty and is a factor in many situations, such as determining insurance rates, making predictions, and studying risk.

Enjoy helping your child begin to explore chance.

Sincerely,

The Mathematics in Context Development Team

Dear Student,

You are about to begin the study of the *Mathematics in Context* unit *Take a Chance*. Think about the following words and what they mean to you: *fair, sure, uncertain, not likely, impossible.* In this unit, you will see how these words are used in mathematics.

You will toss coins and number cubes and record the outcomes. Do you think you can predict how many times a head will come up if you toss a coin a certain number of times? Is the chance of getting a head greater than the chance of getting a tail? As you investigate these ideas, you are beginning the study of probability.

When several different things can happen, you will learn how to count all of the possibilities in a "smart" way. Keep alert during the next few weeks for statements that you may read or hear, such as "The chance of rain is 50%." You might even keep a record of such statements and bring them to share with the class.

We hope you enjoy learning about chance!

Sincerely,

The Mathematics in Context Development Team

1. Put a check in the column that best describes your confidence
that the event will take place.

	Statement	Sure It Won't	Not Sure	Sure It Will
A.	You will have a test in math sometime this year.			
B.	It will rain in your town sometime in the next four days.			
C.	The number of students in your class who can roll their tongues will equal the number of students who cannot.			
D.	You will roll a "7" with a normal number cube.			
E.	In a room of 367 people, two people will have the same birthday.			
F.	New Year's Day will come on the third Monday in January.			
G.	When you toss a coin once, heads will come up.			
H.	If you enter "2 + 2 =" on your calculator, the result will be 4.			

Name_____

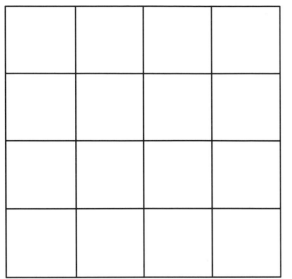

First Floor

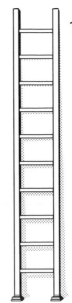

100%

10. a. Color the first floor so that Newton will have a 50% chance of landing on a black square.

b. Mark the 50% chance on the ladder.

c. What is another way of saying: "The Chance is 50%?"

0%

12. a. Color the second floor so that Newton's chance of landing on a black square is 1 out of 5.

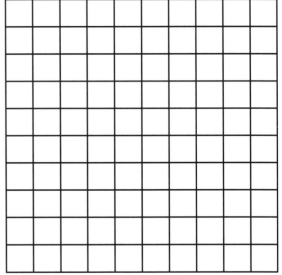

Second Floor

b. Color the third floor with any pattern of black-and-white tiles. What is the chance that Newton will land on a black tile on the floor you made?

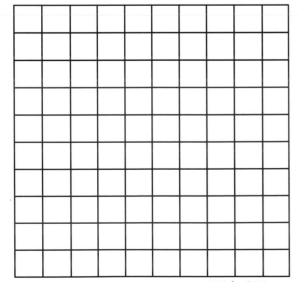

Third Floor

19. Connect all statements that say the same thing.

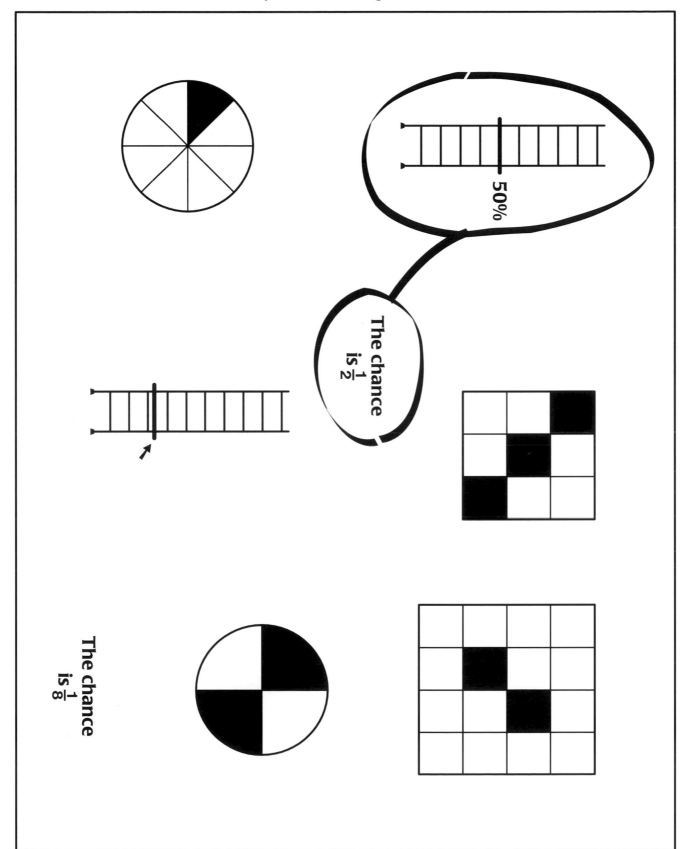

Use with *Take a Chance,* page 32.

13. **a.** Put an H on the tree diagram at the place Harry would start.

   b. If Harry finds the food in room 3, trace Harry's path on the maze.

   c. Trace Harry's path on the tree diagram.

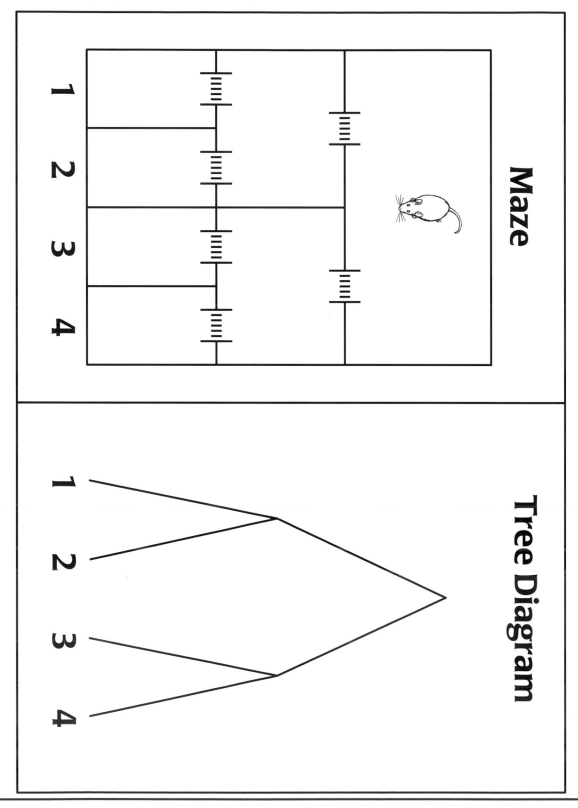

Maze

Tree Diagram

25. **a.** For each square, fill in the sum of the numbers.

b. How many different pairs are possible, when you roll two number cubes?

c. How many ways can you get a sum of 10 with two number cubes.

d. What is the best number to pick if you are playing Sum It Up? Is you answer different from your choice in problem **24**?

PROBLEMS ABOUT CHANCE

Use additional paper as needed.

1. The weatherman announced,

> **THERE IS A 10% CHANCE THAT IT WILL BE DRY ALL DAY TODAY.**

a. Rewrite the above statement using terms such as *likely, sure, very, not, almost,* and so forth.

b. Using the above statement, what do you know about the chance it will *rain* today?

2. Next to each story below, write the number of the chance statement that matches it. Some stories may have more than one matching statement, and some statements may match several stories.

Story

a.  My friend raffled off a prize at a party. He made 15 tickets and gave me three. What are my chances of winning the prize? _____

b. What is the chance that a frog will land on a black square on the floor on the left? _____

c. I am playing a game. If I throw a 3 or a 5 with a number cube, I will win. Is it likely that I will win? _____

d.  Sam is thinking of a number between 1 and 15. What are my chances of guessing his number? _____

e. I throw two coins. What is the chance that I will get a head and a tail? _____

Statement

1. There is a 20% chance that this will happen.

2. There is about a 30% chance.

3. It is not very likely.

4. There is an 80% chance.

5. The chances are 1 out of 3.

6. There is a 50-50 chance.

7. The chances are 1 out of 5.

8. The chance is 0%.

PROBLEMS ABOUT CHANCE

Use additional paper as needed.

3. Katrina is a student in a small school that has only six classes. Two classes at her school are invited to visit a television studio in a nearby city. The principal wants to use a fair method to select which two classes will go.

a. Describe two fair methods the principal can use to make the choice. Why do you think your methods are fair?

Mr. Harris's and Ms. Lyne's classes were picked. Katrina is in Ms. Lyne's class. She wonders what to wear. She takes all of her blouses and slacks out of the closet to see what outfits she might wear. She finds that she has 12 different combinations of a blouse and a pair of slacks.

b. Describe or draw the clothes Katrina might have taken out of her closet.

In the television studio, four of the students will be chosen to be on camera. Mr. Harris's class has 23 students, 8 boys and 15 girls. Ms. Lyne's class has 27 students, 12 boys and 15 girls.

c. Describe how you would pick the four students. Explain why you would use your method.

d. Katrina is in Ms. Lyne's class. She wonders how good her chances are for being chosen. Using your method from part **c,** write her a note describing her chances.

PROBLEMS ABOUT CHANCE

Use additional paper as needed.

4. **a.** Design two different mazes, each with eight rooms, so that each room has an equal chance of being reached.

b. Four different tree diagrams of mazes are shown below. Order the diagrams according to the chance a mouse has of ending up in room 3. Describe how you determined the order.

Maze i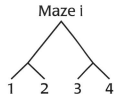

Maze ii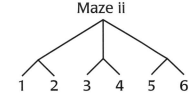

Maze iii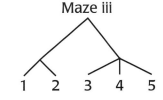

Maze iv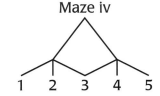

c. Design a maze with eight rooms (different from those you created in part **a**) so that the chance of reaching room 5 is greater than the chance of reaching room 1. Find the chance of reaching room 1 and the chance of reaching room 5 in this maze.

5. Belinda found an old coin. She tossed the coin and heads came up on the first three tosses. What can you say about the coin?

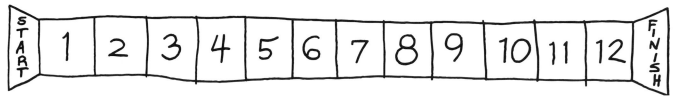

PROBLEMS ABOUT CHANCE

Use additional paper as needed.

6. Paula and Huong made a board game for two players.

| START | 1 | 2 | 3 | 4 | 5 | 6 | 7 | 8 | 9 | 10 | 11 | 12 | FINISH |

A sketch of their board is above. Below are the rules for the game which describe the four possible moves:

> Move i: Go one step forward.
> Move ii: Go three steps forward.
> Move iii: Go three steps backward.
> Move iv: Stay where you are.

They also decided to make some rules about _chance_. There must be a big chance for move i, equal chances for moves ii and iii, and a 25% chance for move iv.

They make a spinner to decide what move to make on each turn. The person who reaches Finish first is the winner.

   a. Design a spinner that fits the rules Paula and Huong made.

   b. Estimate the chance for each of the four moves, either as a fraction or as a percent.

After playing the game, Paula and Huong thought it took too long for the game to end. They decided to change the rules.

   c. How would you change the rules to make the game go faster? Explain your answer.

Section A. Fair

1. **a.** Answers will vary. Refer to the glossary on page 98 of this Teacher Guide.

   b. Answers will vary. Refer to the glossary on page 98 of this Teacher Guide.

2. Answers will vary. Sample responses:

   a. Two teams are playing against each other in a soccer game. They flip a quarter to decide which team gets the ball first.

   b. You say to your friend, "Let's flip a coin to decide who goes first. Heads, I win, and tails, you lose."

3. Explanations will vary. The statement means that, given the exact conditions of the weather on that specific day, if there were 100 days with those exact weather conditions, then it would snow on 70 out of 100 of those days.

   Note: Some students may mistakenly think that a 70% chance of snow means that it is expected to snow for 70% of the day.

4. No, this is not a fair decision. It favors students with brown eyes.

5. **a.** 5 out of 210, or $\frac{5}{210}$, or $\frac{1}{42}$

   b. 3 out of 210, or $\frac{3}{210}$, or $\frac{1}{70}$

   c. You cannot tell from the information given. It is fair if each ticket is sold at the same price and if each ticket has an equal chance of being selected.

6. **a.** This event is not fair. It favors the boys.

   b. This event is not fair. It favors children with brown hair.

   c. This event is fair. Your chances of grabbing a white marble are just as good as your chances of grabbing a red one.

   d. This event is not fair. It is much more likely that you will grab a white marble than a green one.

Section B. What's the Chance?

1.

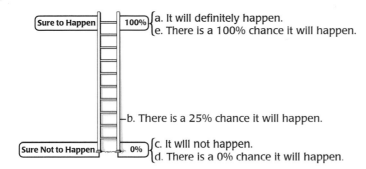

a. It will definitely happen.
e. There is a 100% chance it will happen.

b. There is a 25% chance it will happen.

c. It will not happen.
d. There is a 0% chance it will happen.

2. 1 or 100%

3. 0 or 0%

4. **a.** $\frac{50}{100}$ or $\frac{1}{2}$

   b. $\frac{25}{100}$ or $\frac{1}{4}$

   c. $\frac{75}{100}$ or $\frac{3}{4}$

5. **a.** Drawings will vary. Sample drawing:

   b. Explanations will vary, but should indicate why the drawing in problem **5a** represents a fair spinner. Sample explanation:

   The spinner is half white and half black. So, when you spin it, you can expect it to stop on black half of the time and on white the other half.

6. Drawings will vary. Sample drawings are provided below.

a. 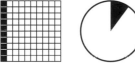   **b.**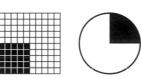

c. 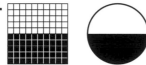   **d.**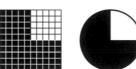

e.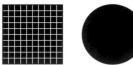

Section C. Let the Good Times Roll

1. **a.** 3 out of 6, $\frac{3}{6}$, $\frac{1}{2}$, or 50%

   b. 3 out of 6, $\frac{3}{6}$, $\frac{1}{2}$, or 50%

   c. 3 out of 6, $\frac{3}{6}$, $\frac{1}{2}$, or 50%

   d. 1 out of 6, or $\frac{1}{6}$

2. If the coin is fair, it should land on heads half of the time and on tails the other half, or around 50 times each.

3. six times

4. **a.** Results will vary. Sample results:

Outcome	Number of Times It Came Up
1	ⲐⲐⲎⲎ
2	ⲐⲎⲎⲎ ‖
3	ⲐⲎⲎⲎ ‖
4	‖‖‖
5	ⲐⲎⲎⲎ ‖‖‖‖
6	ⲐⲎⲎⲎ

   b. Results will vary.

   c. Answers will vary.

   d. Answers will vary. Students should explain variations in their results.

5. 1 out of 6. Each roll of the number cube is independent of the roll before it.

6. No. Each lottery drawing is independent of other lottery drawings. Therefore, if you were to win one drawing, this would not affect your chances of winning another drawing.

Section D. Let Me Count the Ways

1. Answers will vary. Refer to the glossary on page 98 of this Teacher Guide.

2. There are nine possible outfits.
   Drawings will vary, but should show three different shirts and three different pairs of pants.
   Sample drawing:

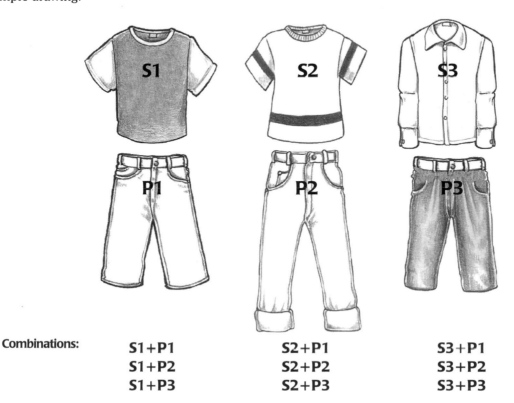

Combinations:	S1+P1	S2+P1	S3+P1
	S1+P2	S2+P2	S3+P2
	S1+P3	S2+P3	S3+P3

3. **a.** 6 out of 36, or $\frac{1}{6}$

   b. 18 out of 36, or $\frac{1}{2}$

   c. 18 out of 36, or $\frac{1}{2}$

   d. 5 out of 36, or $\frac{5}{36}$

   e. 5 out of 36, or $\frac{5}{36}$

4. **a.** 25

   b. 25 out of 100, $\frac{25}{100}$, $\frac{1}{4}$, or 25%

   c. 1 out of 2, $\frac{1}{2}$, or 50%. The chance that a mouse will end up in room 1 is 25%. The chance a mouse will end up in room 2 is 25%. Therefore, the chance a mouse will end up in either room 1 or room 2 is 50%.

5. **a.** Answers will vary. Sample response:

   Since there are more green jelly beans than any other color, I would expect to get a green jelly bean.

   b. 5 out of 50, $\frac{5}{50}$, or $\frac{1}{10}$

CREDITS

Cover

Design by Ralph Paquet/Encyclopædia Britannica Educational Corporation.

Collage by Koorosh Jamalpur/KJ Graphics.

Title Page

Photograph by Robert Drea.

Illustration by Phil Geib/Encyclopædia Britannica Educational Corporation.

Illustrations

6 David Alexovich/Encyclopædia Britannica Educational Corporation; **8** Phil Geib/Encyclopædia Britannica Educational Corporation; **10** Paul Tucker/Encyclopædia Britannica Educational Corporation; **12, 14, 16** Phil Geib/Encyclopædia Britannica Educational Corporation; **18** David Alexovich/ Encyclopædia Britannica Educational Corporation; **24 (top)** Paul Tucker/Encyclopædia Britannica Educational Corporation; **24 (bottom), 28, 30 (bottom)** Phil Geib/Encyclopædia Britannica Educational Corporation; **30 (top)** Paul Tucker/Encyclopædia Britannica Educational Corporation; **32** Jerome Gordon; **34** David Alexovich/Encyclopædia Britannica Educational Corporation; **38** Jerome Gordon; **40** David Alexovich/Encyclopædia Britannica Educational Corporation; **44** Jerome Gordon; **50, 52** Phil Geib/Encyclopædia Britannica Educational Corporation; **54** Paul Tucker/Encyclopædia Britannica Educational Corporation; **58, 64, 66, 68, 70, 72, 74, 76, 78, 80, 82, 84** Phil Geib/Encyclopædia Britannica Educational Corporation; **90, 92, 94, 96, 101** Phil Geib/Encyclopædia Britannica Educational Corporation.

Photographs

6 © Robert Drea; **20 (top)** © Robert Drea; **20 (bottom)** © Ezz Westphal/Encyclopædia Britannica Educational Corporation; **22 (top)** © Joseph Nettis/Tony Stone Images; **22 (bottom, left and right)** © Robert E. Daemmrich/Tony Stone Images; **28** © Robert Drea; **56, 58** © Robert Drea.

Mathematics in Context is a registered trademark of Encyclopædia Britannica Educational Corporation. Other trademarks are registered trademarks of their respective owners.